FEB   2008

THE GOD MACHINE

Bantam Books

THE GOD MACHINE

FROM BOOMERANGS TO BLACK HAWKS: THE STORY OF THE HELICOPTER

JAMES R. CHILES

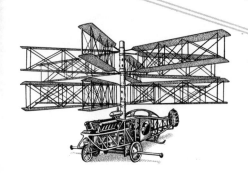

THE GOD MACHINE
A Bantam Book / November 2007

Published by
Bantam Dell
A Division of Random House, Inc.
New York, New York

Book design by Glen Edelstein
Illustrations by Laura Maestro

Library of Congress Cataloging-in-Publication Data
Chiles, James R.
The god machine : from boomerangs to black hawks, the story of the
helicopter / James R. Chiles.
p. cm.
Includes bibliographical references and index.
ISBN 978-0-553-80447-8 (hardcover)
1. Helicopters—History. I. Title.

TL716.C53 2007
629.133'352—dc22
2007028575

Printed in the United States of America
Published simultaneously in Canada

www.bantamdell.com

10 9 8 7 6 5 4 3 2 1
BVG

To Lee Kennedy and the men of his jolly green ship.
Who once set course for Black River,
And passed through fog and a sort of hail,
To bring a man a chair.

Acknowledgments

To my family: Chris, Ben, Jeff, and Kevin for their deep well of patience, and for supporting the project in all respects. One day we will fly together.

A snappy salute to my editor, John Flicker, who as a scout with the Eighty-Second Airborne has had his boots in the air as well as on the ground.

To my agent at William Morris, Jay Mandel, who over more than a year kept encouraging that "Helo" proposal to shape up and ship out.

To historians and helicopter experts, among them E. K. Liberatore, Marat Tishchenko, Jean Boulet, Bruce Charnov, Pete Gillies, Wayne Brown, Lester Grau, Jay Spenser, Michael Lynn, Patrice Bret, Roger Connor, Gordon Leishman, Ray Prouty, and Shawn Coyle.

To Dan Rudert, the first to open up the subculture of professional helicopter pilots for me.

To those who knew Igor Sikorsky—son Sergei, Harry Nachlin, Bob Kretvix, and others—and who told me of his life and times.

To Benjamin Harrison, Hank Emerson and his Recondos, army aviators, and civilians who lived through the Vietnam experience and who passed along their recollections.

For helicopter wranglers at the U.S. Bureau of Land Management, Haverfield, Angel City Air, and the 160th SOAR for giving me a chance to see a day in the life.

And finally, to Link Luckett, John Miller, and all Great Sticks: may your tribe increase.

CONTENTS

THE GOD MACHINE

Introduction

"Why the Birds Sing"

Late in 1909, Wilbur Wright was in the process of training two army lieutenants to fly at College Park, Maryland. He took time on the morning of October 27 to offer a ride to a friend of his sister's, Sarah Van Deman. It would be a historic opportunity, because no woman had yet flown from U.S. soil. Wilbur had already turned down a similar request from President Teddy Roosevelt's daughter. After the four-minute flight, and after untying the twine around her ankles that had kept her skirts from billowing, Wilbur's passenger alighted primly and told friends, family, and Army Signal Corps men, "It was delicious. Now I know why the birds sing."

A little more than three decades later, pioneer helicopter designer Igor Sikorsky predicted in the *Atlantic Monthly* that helicopters "will destroy space for millions of people." The phrase may have sounded ominous, but Igor was a man of peace and he only meant that distance would become irrelevant: in time almost anybody would be able to go about anywhere and helicopters would play an important part in that liberation. Throngs would work in the city, and at eventide would disband to homes in the country, some riding helicopter buses and perhaps a million families using private helicopters. The family copter could even have a balcony on the front. That way all those on their way to the countryside could enjoy the kind of breathtaking view that only helicopter front-seaters enjoy. The industrial and the pastoral would coexist in rotary-winged harmony. As Sarah Van Deman had phrased it, helicopter lovers—I call them *helicoptrians*—would learn why the birds sing.

We know now that a million families never went out and put money down for a helicopter; only a few thousand families have ever owned such a pleasure craft since the first private helicopter sold in 1946. Still, just about anyone with the money can rent a seat and ride along.

And sometimes we can thumb a ride, to a place like Alaska's Lookout Ridge. Well north of the Arctic Circle, this silent, wind-burnished ridge is topped with orderly polygonal slabs of shale and tiny lichens. There are no trees, and little to block the view except more such ridges and rilles. Looking north, feeling the chilly June wind off Alaska's vast northwest shoulder and with the cloud-shrouded foothills of the Brooks Range brooding behind us, I feel safe in saying that this JetRanger turbine helicopter has gotten the three of us rather far into Sikorsky's wilderness. It's a hundred miles from here to our base camp at Ivotuk and the camp is itself another two hundred miles from the nearest paved road. Twenty-three years earlier, I had spent a full week hiking and rafting a hundred miles through the Brooks Range. Today a hundred miles took one hour of flight time.

The helicopter costs the Bureau of Land Management $355 per hour in the air, plus a flat fee of $2,000 per day. That's not counting the sky-high cost of fuel, which has to be flown up in bladders or barrels from far away. Technician Rachael Shields has come here to set up a small radio transmitter that will boost signals from distant archaeological field crews. Pilot Mel Campbell, Vietnam veteran and native of small-town Nebraska, removes a scuffed flight helmet with *Mad Mel* painted on the front and lugs bags and boxes from the cargo compartment to help. Sightseeing is not on the mission orders, but nobody's stopping me from admiring the way that sunlight sweeps across the lowlands, which are striped with snow and bare patches of gravel. Caribou are closing in from the east and west. Other than we three, a heap of electronics, and the magic carpet sitting on a rocky outcropping, there is nothing in sight indicating that humans have left a mark on the Earth.

As the two raise the antenna mast and secure it with white ropes and stakes hammered into the shaly ground, the two rotor blades of the Bell flop in the cold wind. The turbine ticks softly as it cools. I hear frustrated remarks from the pair. Something is balking deep within the circuit boards of the repeater box and Shields cannot raise anyone on the North Slope using her plug-in radio. We load the tools into what passes for a trunk in a helicopter and climb in. Mel starts the turbine and spins up the rotor in a scrupulous sequence, using a digital timer on the instrument panel to let the turbine warm up, though he has started turbines thousands of times. We lift off and head south for camp. I ask him to explain what he is doing along the way, over the intercom. I will be training over the summer—learning why the birds sing and how helicopters do what they do, and want to vacuum up as much wisdom as possible.

While a helicopter can leap over lakes and mountains in a single bound delivering supplies or assistance, this happens mostly out of the public eye. The pro-helicopter crowd feels that the public didn't appreciate the helping hand of helicopters until Hurricane Katrina, and will forget too soon that 200 private and government helicopters swarmed into southern Louisiana to lift 35,500 people from flooded roofs and streets in an effort that was heartwarming but also, at least during the early stages, fraught with danger from midair collisions.

"I've never witnessed anything like it," military pilot Mike Chapman told the *New Orleans Times-Picayune*. "It was the busiest helicopter operation I've ever seen in the world, including Vietnam, including Iraq." Helicopters were so critical, and ordinary affairs so upended, that for a few weeks their operators enjoyed a holiday from the kind of noise-driven public protest that had surrounded an effort earlier that year by New Orleans' Ochsner Hospital to relocate its heliport to a residential area.

Well before Hurricane Katrina, the willingness to reach down from helicopters with that god-like grip included many other volunteer pilots. On his own initiative, news-copter pilot Bob Pettee rescued people during a 1992 flood in Southern California. Afterward he told the *Los Angeles Times,* "I was in Vietnam and I don't need any hero's song for this. There's not a helicopter pilot anywhere that is not spring-loaded* to help somebody." Which is not to say that all such impulses turn out well. In an incident the same year, private helicopter owner Harold Arbeitman crashed while trying to reach an injured boater on the Mississippi River. Safety experts call these White Knight crashes.**

Helicopters allow rescues once unthinkable, such as a December 1994 mission that required two Pave Hawk rescue helicopters to fly far from Nova Scotia, into a storm, after a distress call from the sinking cargo freighter *Salvador Allende.* No other aircraft or ship offered a timely chance of retrieving sailors from the fifty-foot waves of the North Atlantic. This fifteen-hour record-setting mission, reaching out 740 miles from shore and requiring ten midair refuelings (eight of them considered "critical"), recovered one survivor.

Sometimes helicopters go to sea with something else in mind. Late at night in late October and 100 miles north of Cartagena, Colombia, a fiberglass-hulled boat of the type known to international drug-busters as the "go-fast" was crashing through waves at speeds approaching fifty miles per hour. The three 200-horsepower Yamaha outboards could have pushed the boat when empty well past sixty miles per hour, but this one was loaded with nearly a thousand gallons of gasoline and 4,300 pounds of pure cocaine in mailbag-sized bales. The tropical night was a perfect cloaking device: moonless, thickly overcast, and therefore starved even of starlight. The five smugglers on board looked like a sure bet to deliver their $140 million cargo to a lawless stretch of Guatemalan coast for the transfer to truck or air transport. Such boats had been making runs out of Colombia with impunity through 1999. But this was 2004, and the boat's position had already been picked up by distant radar and relayed to the cutter *Gallatin.* Now a deep blue light appeared off to the side of the boat, illuminating a U.S. Coast Guard logo on an Agusta-Westland

* This expression among Vietnam-era helicopter pilots refers to a spring-loaded switch and warning horn on "Huey" helicopters indicating the main rotor was turning too slowly.

** In one particularly fateful turn of events, Senator John Heinz of Pennsylvania was killed in a collision after a helicopter flew close underneath his airplane. It was an attempt by the pilot to determine whether the Heinz airplane's landing gear was down and locked in position.

MH-68 helicopter. The helicopter's crew sent commands via bull-horn and radio, in English and Spanish, for the boat to heave to.

It didn't stop. Using authority vouchsafed to him by radio and crouching in the open left-side doorway, gunner Jason Murphy fired warning shots from a light machine gun, placing the burst well ahead of the boat. The captain didn't respond to that Morse-like message either, so Murphy picked up his next option: a gray, heavy bolt-action rifle. This was a Robar sniper rifle, firing standard .50-caliber rounds identical to those used by heavy machine guns. Murphy took aim at the engine block of one of the outboards. After Murphy's first shot, the boat's captain put the boat into a tight circle to throw off his aim. The helicopter followed the boat closely, at one point settling toward the ocean when it got caught in the downwash of its own rotor blades. The helicopter backed off. The helicopter's fuel exhaustion, or some other in-flight emergency, was the go-fast's only hope of escape now.

It didn't happen. With nine deliberate shots, using a laser sight to fix his aim on the outboards despite the spray shooting up around the stern, Murphy destroyed each engine in turn. Each direct hit, causing a gusher of hot engine oil, showed up clearly on the side-looking infrared screen. The helo-sniper technique combining radar detection, helicopters, rifles, and a boat with a boarding party has worked every time, so far. At last report, the street value of all drugs intercepted this way was $8.5 billion.

Because of the aplomb with which helicopters pull people out of danger and strike down enemies near and far, they have become a real-life match for the Greek theatrical gadget known as the mechane.

No ancient mechane has survived the ravages of time, but we know from drawings that it was a counterweighted, pivoting crane whose tip went up and down when muscle power was applied at the other end. It was adapted from a device called the *geranos,* which Greek builders used for raising stone and wood. The mechane's lifting arm was probably twelve feet long when first used at the theater of Dionysos Eleuthereus in Athens. Historians believe that the first play to employ it was authored by Aeschylus, a war hero among the Greeks for his role at the battle of Marathon. Precisely which production saw the first appearance of the special effect is unknown,[1] but it could have been *Prometheus Bound* or *Psychostasia.*

The subsequent tragedian Euripides locked on to the mechane as a plot-changer also, even more so than had Aeschylus. In Euripides' play *Medea*, the title character sets up a vengeful scheme to poison the new love interest of her husband (Jason, of the Argonauts) and then goes on to murder their own children in Corinth to torment him further. As Jason takes the stage to confront his wife, she appears overhead in a chariot marked with the sign of the god named Helios. Her chariot takes off for Athens, carrying away the corpses of their children as she gloats over Jason's grief. The gullible among the ancient audience may have believed that the sun god Helios really had come down to Earth and was assisting the stage crew, but the stunt was due to the counterbalanced crane arm of the mechane.

Since the English word *machine* derives directly from the Greek term, this theatrical gadget is arguably the mother of machines. Roman playwrights embraced the device also and gave us the Latin phrase *deus ex machina,* meaning "god from a machine." In fiction-writing parlance today it means any last-minute resolution or rescue.

Helicopters are the modern-day mechane. To be a god means to wield power, to change fates, to see into lives, and to have freedom of action. Helicopters facilitate all of these.

The biggest helicopters, which are the size and weight of semi-trailer trucks, have pulled oceangoing freighters out of trouble. A chartered Russian Mi-26,[2] the biggest rotorcraft in action, has gone into the mountains of Afghanistan to haul back a downed CH-47 Chinook.

A great deal of work was required to harness such energies while preserving the subtle touch that the controls on any helicopter must offer a pilot. Proof of difficulty is that 140 years passed from the first spring-powered model helicopters until the flight of a human-carrying helicopter. It's ironic that the gestation of helicopters was so very slow compared to airplanes, because a pair of rotating wings was mankind's first flying machine. The earliest example is an object unearthed at Oblazowa Park in present-day Poland. Radioactive carbon testing of the two-foot-long object, which was carved out of mammoth tusk, indicates that it was crafted 23,000 years ago. While the average citizen seeing the image of the Polish artifact might think "That's a boomerang," it was most likely a hunter's weapon, thrown straight out to kill birds and small animals. To experts this device was a "throwing stick." A boomerang, by contrast, is a recreational device that is shaped and weighted to follow a long looping path. If thrown overhand with sufficient skill, or luck, a boomerang will circle back to land peacefully at the thrower's feet. If not, it can loop back energetically and whack the thrower on the head, even as he turns to run for shelter. Henry A. Wallace, initially secretary of agriculture to Franklin D. Roosevelt and later his vice president, loved boomerangs. Once he broke several fingers while catching one on the return flight.

Multiple ancient cultures independently created throwing sticks or boomerangs, including Egyptians, Paleolithic Dutch and Austrians, and Native Americans. The aborigines of Australia took the ancient technology to its highest level. According to the nineteenth-century missionary Henry Haygarth, a skilled boomerang-wielder could consistently hurl one around a house without it touching the walls.

Throwing sticks and boomerangs all possess wing-shaped ends, which generate a finely balanced lift as the device spins around its center of mass. Inventors of the boomerang may have been inspired by seeds of the maple tree. Maple seeds prove that a rotating wing can descend to a gentle landing, and more reliably than any parachute-like device. Similarly, all helicopters are designed around this "autorotation" principle, so they can flutter to a landing even after the engine fails.

It is striking that for a field so ruled by unforgiving physics, a distinct streak of quirkiness, even obsession, runs through the biographies of many helicopter pioneers, such as V. V. Tatarinov, who in 1909 received money from the Russian War Ministry to build a helicopter. His craft had the shape of a small car with an airplane propeller on the front, topped by a crane-like affair that supported four lifting propellers. Upon looking the machine over, Russian war minister Vladimir Sukhomlinov decided to cut Tatarinov loose. A year later, Tatarinov—driven to "psychic disturbance" by criticism from the press—burned down his laboratory and the Aeromobile along with it.

Another who dreamed powerfully of helicopters was Irish-born inventor Louis Brennan. Even though his prototype helicopter of 1925 crashed during official British tests and eventually proved unable to keep up with advances by others, his story is relevant because he was keenly interested in the strange physics of rotating machinery—much more so than most helicopter tinkerers.

Brennan's middle initial might as well have been "G," for Gyroscope, since rotating toys captivated him as a child and they never lost their allure. Early in his career Brennan sold a gyroscopically stabilized torpedo design to the British Royal Navy. It made him financially secure. As a gentleman inventor so typical of the age, Brennan went on to invent the Gyro-Car, a self-propelled railroad carriage that wowed onlookers during trials in England by balancing and running along a single rail. Whether moving or at rest, Brennan's Gyro-Car was held upright by the cunningly exploited gyroscopic action of two massive, high-speed flywheels.* In one demonstration at his estate he placed his small daughter on a five-foot-long model and sent her out on a tightrope suspended over a small valley to demonstrate its poise.

Brennan also predicted that giant fortress-helicopters would one day shield the great cities of Britain from bomber attack. Such a rotationally obsessed person deserves honorable mention in a helicopter history because the single most important principle in understanding the helicopter's behavior, which can seem bizarre if not controlled, is rotation. Hanging a fuselage under large, spinning blades allows humans to hover, but it's only safe if someone pays close attention. Without special provision, helicopters are unstable when hovering and therefore take much time to master. When a helicopter is hovering under an experienced pilot's fine control, parked in the air as if held by ropes, it may appear fully subdued but in fact is carrying a great deal of energy.

In August 1948 aeronautics expert Alexander Klemin published an article in the *Scientific Monthly* titled "The Helicopter Problem." The purpose of his article was to explain why the craft had not yet taken to the skies in great numbers now that the war was over. His title is equally descriptive of the entire subject's history. Experimenters spent many decades getting progressively larger helicopters to behave themselves in the air. They tackled early problems and solved them, but each resolution allowed a new set of difficulties to take center

---

* In one trial, 107 passengers crowded aboard the Gyro-Car. Brennan predicted that hotel-sized versions of such vehicles would cruise at 150 miles per hour between California and New York, vaulting canyons and rivers on single-rail trestles.

stage. Hovering an early helicopter off the ground for more than a few seconds was reason to celebrate, but freeing that aircraft from the surly bonds of earth allowed severe vibrations to take over. Over time this string of challenges changed from technical ones to tactical ones, and even to those of public relations. In that spirit, the sixteen chapters of this book are organized around a specific helicopter problem and its resolution. These are followed by a timeline of helicopter milestones.

The first chapter, "Preflight," is an introduction to the working parts of the machine. The next two chapters cover the prehistory: the working models, balloons, gliders, and engines, along with the dreamers who laid the foundations. Chapter Four, "Breakout," describes how experiments and understanding came together in the 1920s to yield workable if clunky machines. Then helicopters seemed to lose their momentum to a radically new rotorcraft from Spain ("Automobile of the Air"). At long last, travel-worthy helicopters arrived in the late 1930s, culminating in a streamlined design by Igor Sikorsky ("Jules Verne in a Suitcase"). Chapter Seven profiles inventors who began manufacturing true helicopters during the war, and Chapter Eight ("The Race") describes how all those makers planned to churn up a huge demand and then satisfy it. As it turned out, the public loved to read stories about helicopters but didn't line up to buy them. By late 1949 the industry was in a downward slide.

The following year the Korean War opened, and solved the problem of demand. Almost any production helicopter, even those with two or three seats, was bought in quantity for testing or, in some cases, for immediate use. Chapter Nine ("The Tyrant") describes how the marines greatly expanded the usefulness of helicopters on the battlefield by delivering troops and supplies in the heat of battle. Helicopters offered commanders a chance to rise above the fog of war. "Operation Chopper" explains how in Vietnam the American military tried to apply lessons from Malaya and Algeria. Before the United States cut back its war effort, certain infantry units, such as the forces led by Colonel Henry Emerson, discovered new ways for helicopters to meet their combat potential. The following chapter, "Last Man Out," leads to the helicopter's final dramatic role in Vietnam: an escape vehicle for American citizens, their guards, and selected allies.

The chapter "Back in the World" narrates the spread of military-designed helicopters and military-trained pilots into the civilian market following Vietnam. The results were sometimes inspiring and sometimes tragic. Public opposition to heliports grew. "Chariots of the Gods" looks at the high costs of helicopters and the consequences. The next chapter ("Something to Watch Over Me") argues that while some citizens worry about domestic misuse of government helicopters, their alter egos, called news-copters, are providing a way for citizens to keep a watch on government in turn, as those eyes in the sky go about their bread-and-butter work of gathering news about house fires, kidnapper-versus-SWAT standoffs, and car chases. The skills that distinguish the best pilots are the subject of "The Great Stick." The conclusion compares the early predictions of mass helicopter use to the way things turned out.

In the end, chapters in a book are nothing more than stories flying in close formation, and helicopters can't be grasped without spending lots of time with those who fly them. Whole bookshelves could be filled with helicopter technical manuals and manufacturers' pilot handbooks, but much wisdom would still be missing. The culture of helicopter pilots moves via word of mouth: Which is the best helicopter on which to train? Can piston engines be as reliable as turbines? Can helicopters chop out their own clearing in a grove of trees? They recount the story of a big twin-rotor Chinook, an experimental gunship positively bristling with cannons and machine guns, that shot itself down over Vietnam. Of a gunship pilot who landed on his enemies to squash them when he ran out of ammunition. Of a "longline" helicopter pilot who forgot he was trailing a steel cable and cast himself into a river. Of a helicopter that lost power while hunting oil in the mountains and lodged perfectly in the fork of a tree.

The machines that catch our eyes now, so capable and powerful, so liberating to some and so awe-inspiring to others, all grew out of prototypes that, despite decades of trying, did not want to put anybody into the air. The few that rose off the ground tried to shake themselves apart. Inventors learned immediately that the job of transforming these wobbly endangerments into a chariot for gods would be an Olympian proposition.

Chapter One

Preflight

Of all birds, winged mammals, and insects, very few have mastered the skill of pausing in midair and going backward as well as forward, so anything capable of such flight is, ipso facto, a rare beast. The ruby-throated hummingbird, which can hover with sewing-machine-like precision and also fly more than 500 miles across the Gulf of Mexico without a rest, is one such improbability. Helicopters are another unlikelihood. Explaining how the parts work together to do the unlikely is best approached by treating the helicopter as terra incognita, exploring it from the headland of its cabin to the archipelago of its tail boom.

This particular helicopter is white, and composed mostly of high-strength steel and aluminum. It is thirty-one feet long and seats two people, typically a wary instructor on the left and a trainee on the right, but it is also suitable for aerial photography and other daily errands. Anyone renting a Schweizer 300C for weekend travel will be traveling lightly, because there is no trunk for baggage.

I required a formal introduction to the Schweizer, because I would be flying one. The instructor I secured from Hummingbird Aviation, John Lancaster, had been a professional skiing instructor for twenty years in Vail, Colorado. Skiing injuries and a love of flight prompted him to seek out a new, and statistically safer, profession. He learned to fly helicopters in Florida at the world's largest privately run helicopter school[1] and came to Minnesota to share his knowledge. Typical students were those planning to fly for police departments, tourist outfits, or offshore oil companies. He favored shorts and sport shirts and, before getting down to business, displayed the cheery demeanor of a camp counselor. But he also had nearly a thousand hours of helicopter time and we were both mindful that I was not here to interview him; he was now my instructor. As when dealing with any rookie, his first job was to explain important parts of the ship.

Lancaster began my first lesson in the flight-school office by picking up a black-and-white toy police helicopter from a shelf and explaining basic principles of flight. Then he sold me a set of pilot books out of a glass showcase that looked to have been bought from a going-out-of-business sale at a jewelry store.

The helicopter he would train me to fly is of recent vintage but a lineal descendant of a 1956 model invented by the Aircraft Division of Hughes Tool. Beginning in 1964, the army used a military version called the TH-55 Osage[*] to drive thousands of new pilots through eighty hours of basic flying at Fort Wolters, Texas. The TH-55 was a light piston-engine helicopter and cost much less than a turbine-powered model to operate.

This session lasted an hour and was devoted to a four-page inspection checklist. It required a close look at every side of this helicopter, including the underside. It was so thorough that we might as well have been hunting for a small bomb that someone had tucked away on the aircraft. With this guidance, Lancaster said, I would know how to check out the machine for my own flights. It seemed to this novice that the whole machine is so small and so open to view

---

[*] It was a trainer version of the Hughes 269. Many pilots called it the "Mattel Messerschmitt." When a small tornado passed over the Downing Heliport at Wolters on April 13, 1967, the storm damaged or destroyed 179 of them.

that it can be taken in with a quick glance followed by a nod or a frown, but this is considered bad form.

Still, an open-minded, undistracted glance from thirty feet away is not a bad way to begin a preflight, Lancaster said. Sometimes it reveals immediately that something is out of whack, or missing, and a mechanic can be summoned sooner rather than later. We folded our arms and took it in, top to bottom. For all the publicity about helicopters black and deadly, this ranked no more than a "1" on the intimidation scale. It has the look of a dragonfly carrying saddlebags, which in this case are twin fuel tanks. It also doesn't look radically more elaborate than a car, so I had to wonder why it costs a quarter-million dollars if purchased new. With an instructor, each hour in the air costs $300 in rental fees. Since helicopter pilots need to fly regularly to stay on top of their skills, winning the license is comparable to buying a fancy boat: the initial expenditure is steep and leads to still more spending.

The Schweizer sits on tubular landing skids that hold it high off the ground, giving it a poised and alert look. Above the cabin is an elaborate-looking hub gripping three airfoils that droop slightly. Each airfoil is a "rotor blade," taken together they make a "rotor." The extreme engineering challenges that are posed by spinning such large weights atop a hovering aircraft came as a complete surprise to

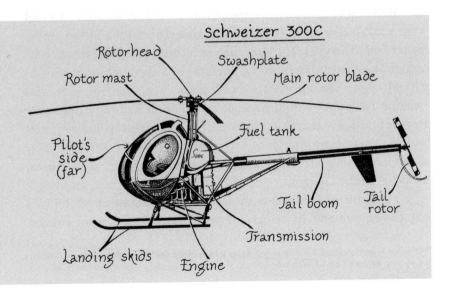

Schweizer 300C

Rotorhead
Rotor mast
Swashplate
Main rotor blade
Pilot's side (far)
Fuel tank
Tail boom
Tail rotor
Transmission
Landing skids
Engine

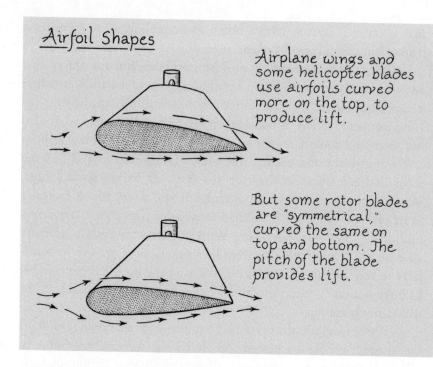

Airfoil Shapes

Airplane wings and some helicopter blades use airfoils curved more on the top, to produce lift.

But some rotor blades are "symmetrical," curved the same on top and bottom. The pitch of the blade provides lift.

the early inventors. When in operation the rotor rotates nearly eight times every second. Unlike airplane wings, which have a more generous curve on the upper side than the lower side, the cross-section of each rotor blade at the end is nearly symmetrical.

Pointing to the rear is a slender, tubular tail boom with a second set of rotor blades at the end, much smaller than the big blades on the main rotor. The tail rotor looks like an afterthought, but it is a technological survivor despite many attempts by helicopter innovators to get rid of it. The tail rotor's side-thrust overcomes the powerful twisting force that would result if a helicopter had only one rotor atop its fuselage and nothing to oppose it.* Single-propeller airplanes would also have a torque problem were it not for the fact that their wings have enough leverage in flight to keep the comparatively small propeller under control.

Lancaster closed in to work the checklist. He squatted to point out the engine and its associated parts, nestled in a stout tubular

* One way to see the effects of uncompensated torque is by firing up a rotary lawn mower in the wintertime and pushing it out onto a sheet of ice. The lawn mower will spin in a direction opposite to that of the rotating blade inside.

frame under the cabin. It's an aircraft engine, built for dependability and costing much more than a car engine. It burns eleven gallons of aviation gasoline per hour.[2] We drained a pint of blue fuel into a plastic jar, checked for water that could kill the engine, then poured it back into the gas tank. Lancaster rapped on the tank, starting high to low, stopping when the sound changed from boom to thump. This verified the actual fuel level. He warned me not to place all my faith in the fuel gauge inside the cabin.

Of all the mistakes helicopter pilots can make, leaving the ground with too little gas for the trip is among the most embarrassing and avoidable. Despite the early freewheeling style of helicoptrians of the 1950s, who felt they could land about anywhere, today it is not acceptable to plan on refilling at a convenience store. That leaves mostly airports, except for the very few cities in the nation offering public heliports. A highly modified helicopter has flown without refueling for fifteen hours, but more typical is two to three hours of operation on a full tank.

We peered into the cabin to make sure the fire extinguisher and paperwork were where they belonged. The view of the outside world is impressive from the pilot's seat, which in helicopters is on the right side of the cabin. The reason is that most pilots are right-handed and this arrangement allows these pilots to control the cyclic lever, which requires the most finesse, with their favored hand. In many conventional airplanes the pilot's visibility is narrow, comparable to looking out from a foxhole with a low roof, but not so with helicopters. The control panel and instruments are kept to a minimum so they all can be packed onto a pedestal in the center. Everything else within sight is a plastic bubble or just open air, since pilots prefer to fly without doors whenever the weather permits. The seat belts are comparable to those in race cars.

Lancaster prompted me to check the engine oil, the transmission bolts, the emergency transmitter, the door hinges, and even the welds in the landing gear. We lit up the beacons and wiped the housings with rags. We wiggled cables, belts, rods, and ball joints. We looked at a frame holding the belts for the transmission; Lancaster recalled a case from the National Transportation Safety Board files where that frame broke and killed the pilot. Unlike gasoline-powered airplanes, where the propeller is bolted directly to the crankshaft, all helicopters must have reduction gears so that the rotor turns only a tenth as fast (or even less) as the engine. The reason is that slow-turning rotors are more efficient than fast rotors.*

"Reading through the NTSB cases makes for an instructive evening," said Lancaster, "but it can dampen your enthusiasm."

We proceeded along the tail boom to the rear of the aircraft. Lancaster stopped for a lengthy discussion about the health and well-being of the tail rotor, which from our first glance at a distance looked to be little more than a set of striped lawn-mower blades. It's not, said Lancaster. There are little rods and bearings that allow the tail-rotor blades to change their angle a little, or a lot, as the pilot steps on the tail-rotor pedals. It has its own gearbox and oil supply.

The tail assembly is so delicate and essential to flight that pilots have to order passengers to be mindful of anything that wind might scour from the cabin, in flight, when the doors are off. Murphy's Law dictates that the slipstream will carry any such object straight into the tail rotor rather than allow it to tumble ineffectually to the ground. While there are verifiable wartime stories of Huey pilots who slashed through branches, even four-inch limbs, with their main rotors on the way down to a landing zone, no sane pilot would plan on chopping or even brushing anything lightly with the tail rotor.

Now fully two pages into the checklist, we reversed course and worked back toward the nose, along the right side of the helicopter. Lancaster pointed out a little window built into the tail boom and prodded it with a finger. It's there for inspection of the tail-rotor driveshaft dampers. These are doughnut-shaped bearings that keep the tail-rotor driveshaft from wobbling, which was one of many problems to vex early helicopters. This brings up a principle applying to helicopter components and helps explain why everything in this avocation costs so much. Let's say Part A is vital to flight. But Part A vibrates if left to its own devices and will crack if neglected, so we need a little Part B nearby, keeping Part A safe. But Part B is dogged by problems of its own, so we need a Part C, and sometimes even a Part D, to guard Part C.

---

* Helicopters will never match airplanes for speed and fuel efficiency. The reasons for helicopters' lower efficiency include air resistance from the rotor mast and the rotor hub, and the fact that the rotor blades can't pass through the air as smoothly and uniformly as airplane wings. And some power must always be wasted as a result of the fact that a portion of every rotor blade spends part of its time flying backward.

Lancaster had me slide my fingers along a skein of steel control cables under the cabin to see if they had unraveled since the last pilot checked them an hour ago. They hadn't. Now it was time to clamber up the rotor mast. Lancaster showed me where to put my feet so I could get a good look at the rotor hub without snapping something by stepping on it.

Sometimes places and machines are said to have a "heart," which in a home could be a cozy kitchen. If helicopters have a heart, something that is unique and important to them in the way that a nuclear reactor is to a ballistic-missile submarine, it would be the rotor hub.

At the high point of this Schweizer, eight feet above the ground, three rotor blades are held in the bolted grip of a metal hub. Each nut has a little loop of safety wire to prevent it from wiggling loose. The white blades have the shape of long, slender wings, which is exactly what they are. The hub not only holds on to the blades despite extreme centripetal forces at full speed, the hub also allows the pilot down in the cabin to alter the angle of each blade in flight with extraordinary delicacy.

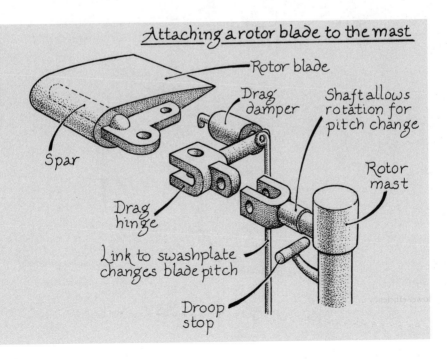

Attaching a rotor blade to the mast

Rotor blade

Drag damper

Shaft allows rotation for pitch change

Spar

Rotor mast

Drag hinge

Link to swashplate changes blade pitch

Droop stop

The hub is arranged so that commands from the cyclic and collective controls can pass through a rotating mechanism called the swashplate and out to the blades. The idea of the swashplate originated with windmill builders in the nineteenth century. If the lower swashplate pushes straight up or down, that changes the collective

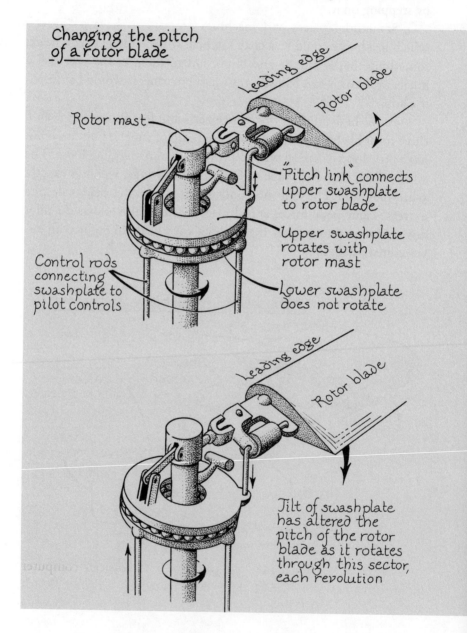

Changing the pitch of a rotor blade

Leading edge

Rotor blade

Rotor mast

"Pitch link" connects upper swashplate to rotor blade

Upper swashplate rotates with rotor mast

Control rods connecting swashplate to pilot controls

Lower swashplate does not rotate

Leading edge

Rotor blade

Tilt of swashplate has altered the pitch of the rotor blade as it rotates through this sector, each revolution

pitch of all the blades. If the lower swashplate tilts to one side, each blade must change pitch while it goes around.

The notion of two plates rotating against each other and forcing the upper one to follow the tilts and pushes of the lower one sounds cumbersome, but in a well-maintained helicopter the linkage is so precise that even imperceptible hand movements of the pilot have a prompt effect.

During flight, the entire loaded weight of the helicopter hangs from the hammer-forged assemblage of parts called the rotor hub. Turbulence or extreme maneuvers multiply that weight by G-forces, such that the hub of the massive Mi-26 cargo copter must be able to tolerate more than 100 tons hanging below. As Igor Sikorsky once said to a reporter who asked him what would happen if the rotor came off in flight: "That is not recommended." The importance of a light, intricate, and yet strong rotor hub is another reason why helicopters cost ten to twenty times more than a car.

There was only room up here for one person to stand and Lancaster called up for me to shove the rotor around by hand, looking for anything wrong with something called the lead-lag dampers. These look like small, black hydraulic cylinders and sit near the base of each rotor blade. These humble devices keep the rotor blades from finding synchrony with the landing gear. Without them the helicopter would be at risk of going into a manic dance and rolling over on its side while just sitting on the ground and idling its rotors. Missing or failed dampers would pose a risk in the air also.

We finished the preflight in a little less than an hour. This seemed like a lot, but never fear, said Lancaster: After I gain more practice in peering, tugging, and wiggling, a thorough inspection at the start of a flying day will take only fifteen minutes. According to veterans, any time spent by a pilot in knowing every part of his or her aircraft will not be wasted. "They say you can teach a monkey to fly, but to know the systems is very important," Alaskan pilot Mel Campbell told me while steering across the North Slope. "If you know what the aircraft will do in all circumstances you can save yourself."

Lancaster escorted me to the school's trailer-sized computer flight simulator, which at $100 per hour costs only a third as much as a real helicopter. We climbed into the miniature cockpit and I buckled my seat belt on the right side. The buckling amused Lancaster because the simulator offers no fancy actuators to make the little cabin rock and roll, as in an amusement ride. It has no more inclination to roll over than does my easy chair at home. The big image of a simulated Chicago O'Hare International Airport was bright and the computer was prepared to make the controls realistically frustrating to the novice. My job, he said, would be to maneuver around a simulated white bus, not hitting anything and not touching the ground.

The three main controls in the simulator's cabin are laid out identically to the real thing. On the floor in front of each seat are two foot pedals, which turn the helicopter left or right. In front of each seat is a joystick coming up from the floor, called a cyclic. The cyclic controls horizontal motion by fine-tuning the pitch of the rotor blades as they go around. This tilts the helicopter slightly, sending it sideways, forward, or backward. In medium and large helicopters pilots would be quickly exhausted by the sheer physical effort of changing the blade pitch, so hydraulic power helps them out, similar to power steering in a car. Some early helicopters such as the

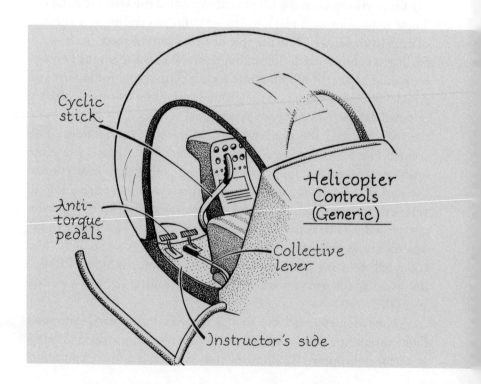

Sikorsky S-51 had long sticks to give the pilot more leverage, for the same reason that big old Buicks and Fords once had big old steering wheels.

At the left side of the seat there is a black lever called the collective. This lever, which looks like a glorified version of the emergency-brake handle found at the side of drivers' seats in some cars, sends the helicopter up or down by changing the pitch of all the main rotor blades at once. A twist-grip throttle, on the end of the collective lever, helps the pilot make sure the rotor's rate of revolution stays within a "green" range. Bad things happen outside this range, particularly if the rotor speed drops too low in flight.

Flying near the computerized Chicago O'Hare went fine until I had to fly slowly and hover. I lost control in just seconds and the helicopter spun, climbed, scooted, and dropped. "Now you're starting to see what you're in for," said Lancaster.

When Lancaster told me to make a wide circle around a bus I backed away from the vehicle unconsciously. This accelerated in an alarming fashion, as if I were caught in a repulsor beam, until I ran tail-first into a terminal building. After several of these incidents the simulator shut down—to cool off, according to Lancaster—but it could have been showing apprehension. I wondered how the real thing would go.

Lancaster said one of the first tasks of the rookie is to learn how the controls work in isolation; how they work together would come later. To begin, it helps to know that the pedals link only to the blades on the tail rotor. The cyclic and collective levers are connected only to the blades on the main rotor overhead. But everything is connected to the same engine; therefore, what happens to one moving surface must affect the others by increasing or decreasing the power available.

The secret of the tail-rotor pedals is easy to explain. Pushing on the left one makes the tail rotor dig deeper into the air. This adds a rightward thrust at the tail, which forces the helicopter nose to the left. Pushing on the right pedal makes the tail rotor ease off, and produces an opposite effect.

The cyclic and collective controls are a bigger challenge to explain, but an analogy helped me remember what they do. Imagine borrowing two canoe paddles. Stand with a canoe paddle in each

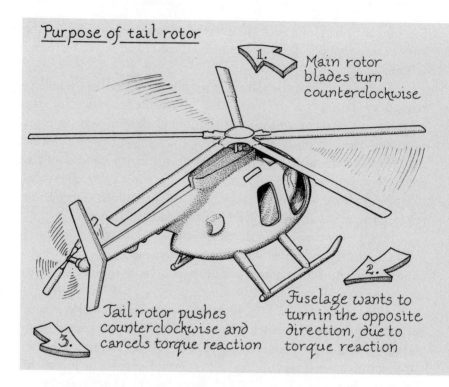

Purpose of tail rotor

1. Main rotor blades turn counterclockwise

2. Fuselage wants to turn in the opposite direction, due to torque reaction

3. Tail rotor pushes counterclockwise and cancels torque reaction

hand, arms extended, and try to hold each paddle straight out, but only by holding the end of the handle. The paddles will droop despite your strongest exertions because no force is helping you hold them up. Those paddles act much like helicopter rotor blades at rest, which would also droop if not supported. Now go to a playground and stand at the center of a merry-go-round. If a friend begins spinning the merry-go-round, you will find that the ends of the canoe paddles rise up, as centripetal force draws them outward. You are imitating a two-bladed helicopter like the Huey.

As the rotation speed increases enough, centripetal force will hold those paddles horizontally. Centripetal force is very important on helicopters too, since it supports the blades despite the weight of the helicopter they are holding up. If for some reason the engine was to produce too little power, the blades would hinge upward at the hub. The playground analogy also shines a light onto what helicopter controls do. On the merry-go-round you could twist your wrists just a little so that paddles bite more steeply into the air, or less steeply. On a helicopter the amount of twist is called blade pitch.

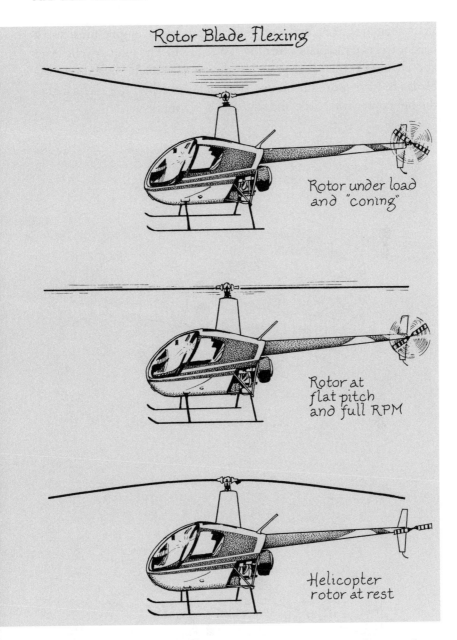

Rotor Blade Flexing

Rotor under load and "coning"

Rotor at flat pitch and full RPM

Helicopter rotor at rest

Twisting both canoe paddles in the same fashion simultaneously—
which has the effect of raising them up or lowering them together—
is what the collective control does for a helicopter pilot. When the
pilot pulls up on the collective stick, the pitch on all blades steepens
and the helicopter goes up. Nothing is free in aerodynamics, and the

extra drag acts like a brake on rotor speed, so the engine must work harder to keep the blades turning fast enough. The pilot's cyclic lever also changes rotor-blade pitch but does it selectively: by causing the blades to have more lift on one side than the other, the cyclic makes the helicopter tilt a little and slide off horizontally.

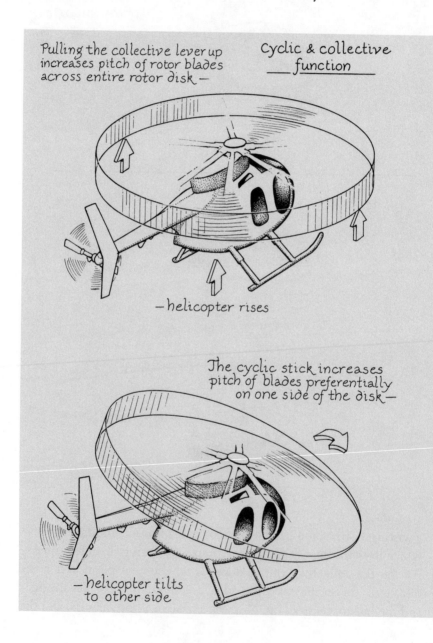

Pulling the collective lever up increases pitch of rotor blades across entire rotor disk —

Cyclic & collective function

— helicopter rises

The cyclic stick increases pitch of blades preferentially on one side of the disk —

— helicopter tilts to other side

THE GOD MACHINE   25

A rescue at the *Ocean Express* drilling rig shows how powerful yet precise these three controls can be with the right hands and feet on board. No other type of aircraft could have pulled it off, no matter how skilled the pilot. And this pilot was very good. Commander John M. Lewis of the U.S. Coast Guard got the distress call at 8:00 P.M. on April 15, 1976, while at Corpus Christi Naval Air Station. The *Ocean Express* had been under tow as a gale came up, but one of the tugs had been disabled and now the rig was listing and about to sink. It was a jackup rig, meaning that it had three giant legs built of steel lattice, something like giant crane booms. The 312-foot-long legs had been raised from the seafloor and now towered high over the rig. These three towers would loom like giant claws for any aircraft attempting to get in close. Lewis flew his Sikorsky HH-52 Sea Guardian, a single-engine helicopter, to the site forty miles east of the Gulf Coast. He made slow orbits, trying to make sense of the scene in the darkness and to calculate his odds of getting in and out. Only one man, Pete Vandicraft, the master of the rig, was still on board. Lewis made two attempts to get a rescue basket within Vandicraft's reach, each time avoiding the three legs, which shifted as the rig slowly rolled over. Salt spray from the storm waves blew into the helicopter. Captain H. B. Thorsen, piloting another Coast Guard helicopter, turned on his spotlight. Vandicraft's helicopter went out of sight in the spray and Thorsen thought the waves had claimed Lewis's Sea Guardian entirely. The scene may have called to mind the legendary credo of the United States Life-Saving Service in the days of wooden lifeboats: "The regulations say you have to go out; they don't say you have to come back."

But the Sea Guardian appeared out of the spray like a surfacing submarine. Witnesses said they had never seen anything like it. Lewis made a third attempt, and came away with Vandicraft in the rescue basket. In departing, Lewis used the controls like so: He pulled the collective lever up to gain altitude, he pushed the tail-rotor pedals to maintain his direction as the rotor blades dug in, and he nudged the cyclic stick to move horizontally away from the wreck. "Piece of cake," as pilots like to say. The rig sank thirty seconds later.

Chapter Two

The Icarian Sea

In 1714 the French writer known as Marivaux gained a measure of fame with a short novel whose title translates as *The Coach Stuck in the Mud*. In the book, five men and women are riding via stagecoach to the city of Nemours. On the way the coachman hops off to do some drinking, which leaves the horses to steer themselves. The coachman wrongly assumes that the horses will manage fine until he catches up on foot. But in time the horses stray off the road in search of grass, which plants the vehicle immovably in mud. Unable to locate their coachman, the passengers walk off to an inn, where they spend all night telling stories.

Marivaux's plot device, along with many such tales of mishap and misery by coach, rang very true for European and British readers through the end of the eighteenth century. Though sailing ships crossed five hundred miles with speed and dispatch, anyone expecting to travel such a distance by land faced weeks of travail, even when taking the express vehicles called flying coaches. Railways and train stations lay years in the future.

The onset of manned balloon flights in 1783, organized by Joseph and Jacques-Étienne Montgolfier, appeared to open another avenue for travel, but balloons moved at the whim of the winds. During a test flight the previous winter, their hot-air balloon had broken from its tether and landed a mile away, terrifying onlookers. Therefore, not long after manned balloons began rising majestically over the cities of Europe and capturing the public imagination in a way no other breakthrough ever had, a handful of experimenters in Europe began thinking about alternatives to the classic balloon.

One of those possibilities was a mechanically powered air-carriage, literally a "flying coach." Experiments in Russia at the middle of the eighteenth century suggested that devices topped by spinning rotors might one day be able to raise themselves off the ground.*

Unfortunately, Mikhail Lomonosov's "aerodromic machine," as demonstrated to the Russian Academy of Sciences in 1754, couldn't do more than dangle on a string and demonstrate a modest amount of lift. The hotheaded[1] Russian had too many other projects underway to perfect his model, but two Frenchmen took up the challenge during the heady months just after the first balloon flights and before the Reign of Terror.

What the two men built was approximately a foot wide, and had two counterrotating rotors, one upper and one lower. Each rotor had two blades made of wood and silk.** It was powered by a small bow, made of whalebone, which when bent stored up energy like a wound steel spring. The energy was delivered to one rotor through a bowstring. The two inventors, named Launoy and Bienvenu, demonstrated their gadget in front of a committee of the French Academy of Sciences on April 28, 1784. A member wrote this in a report filed four days later: "The unbending bow rotates rapidly, the upper wings one way and the lower wings the other way, these wings being arranged so that the horizontal percussions of the air neutralize each other, and the vertical percussions combine to raise the machine. It therefore rises and falls back afterward from its own weight."

* Leonardo da Vinci had sketched a helix-wing hovering machine three centuries earlier, but never attempted to build it.

** The device as often pictured and attributed to Launoy-Bienvenu, with rotors consisting of four blades made of turkey feathers, was actually the one constructed later by George Cayley in 1792.

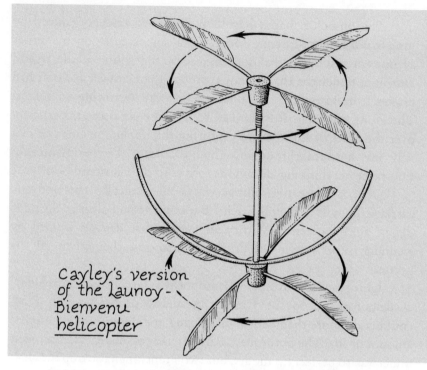

Cayley's version
of the Launoy -
Bienvenu
helicopter

Though not many people paid attention to non-balloons during the excitement of the Balloon Age, it was the first workable helicopter in history, and it also made the first verifiable free flight of any power-driven aircraft.*

The toy-like device of Launoy and Bienvenu was even more important in the way that, years later, it drew certain people into aviation at an early age. Helicopter history books have no more to say about these two other than to list their last names and say that Launoy was a naturalist and his partner was some kind of mechanic. No first names or any biographical details are given in these brief references.

Fortunately, more information than that can be teased from the manuscripts, periodicals, and books of pre-Revolutionary France. "Bienvenu" was the artisan-entrepreneur François Bienvenu, born in 1758. Bienvenu kept a shop at No. 18, Rue de Rohan, in Paris.[2] He was

---

* According to an authoritative study of manuscripts by helicopter historian E. K. Liberatore (1998), there is no evidence that the earlier toy-like devices in China and medieval Europe, later dubbed "flying tops" by some historians, ever flew or were meant to fly. Rather, they were noisemaking toys whose blades produced a humming sound when spun. Liberatore argues that naturally spinning objects such as maple seeds would have provided sufficient inspiration for European rotorcraft work beginning in the eighteenth century.

in the business of making and selling scientific instruments, electrical gear, and gadgets appealing to the public's love of science. It was an immensely exciting time, when people from all economic strata sought to understand, or at least be cocktail-conversant with, the latest discoveries in science and nature. One of Bienvenu's popular products was a suitcase-sized physics-experiment kit targeted at wealthy Parisians who wanted to amuse guests at their country retreats.

For a modest price Bienvenu offered popular lectures on physics, which he convened in his shop as a way to draw floor traffic for his goods. Such courses were extraordinarily popular at this time of the Enlightenment. Other "mountebanks" set up tables and held their scientific demonstrations in the street: for a small charge, some would administer an electric shock, or demonstrate how a jar of hydrogen sulfide gas would reveal invisible writing, or allow a peek through a telescope or a microscope. While most street-side science displays had the look of carnival acts rather than lectures, ordinary people expected some useful information along with the entertainment, even to the point of attending long talks about mathematics.

Considerably less is known about Bienvenu's partner. In all the publications that he is known to be connected with, including a receipt he signed in January 1785, he declined to provide a first name. Nowhere did the helicoptrian Launoy write down a place of residence, a hometown, or any other details of his origin. Though he listed his profession as naturalist, Launoy is not listed in directories covering French naturalists of the era. A newspaper advertisement that Launoy published in 1784 indicated that he could be contacted through the Bureau of Mineral Waters in Paris, an agency that policed mineral-water spas to block false claims and sales of bad water. However, the records of that bureau and its parent organization show no such employee.

The enigmatic experimenter may have been a nom de plume of Claude Jean Veau de Launay (1755–1826), a naturalist and scientist who was active in Paris at the time. There is circumstantial evidence that Veau de Launay had a connection to François Bienvenu: Veau de Launay grew up in the town of Sainte-Maure en Touraine; this was also Bienvenu's hometown. Bienvenu was a builder of electrical instruments; Veau de Launay had a strong interst in electrical instruments and published a catalog that mentioned Bienvenu's gear.

In June 1784 François Bienvenu and the mysterious Launoy arranged a public showing of a device with the redundant name of the Machine Mécanique. The locations were Bienvenu's shop and a coffeehouse in a center of entertainment and shopping known as the Palais Royal. (The Palais Royal was something like a shopping mall for the Ancien Régime, though bawdier. In July 1789, this same neighborhood would see the Revolution ignited, hearing the first infuriated cries for liberty, equality, and fraternity.) At the price of a few coins the two men sold a promotional brochure, which was written to generate backing for a larger, man-carrying model. The sales pitch did not bring in the money required, but the Launoy-Bienvenu helicopter model began making its mark on aviation history by catching the interest of a young man in Yorkshire, England.

His name was George Cayley, born in 1773. Just as the young Wright brothers would later be pulled into aviation by a similar tiny helicopter model, the Launoy-Bienvenu helicopter model inspired this youngster to take a lifelong interest in flying machines. At the age of twenty-three, in two hours of work he assembled a whalebone-bow-powered helicopter styled along their creation. As Orville Wright said, this country squire "knew more of the principles of aeronautics than his predecessors, and as much as any who have followed him up to the end of the nineteenth century."

Cayley had the time to work out principles of unmanned flight, along with many other speculative subjects, because he had inherited a stretch of farmlands in Yorkshire and Lincolnshire, along with the title of baronet, from his father. Others in his position might have dawdled away the pleasant days in comfort, but Cayley devoted his life to the goals of a group of idealists called the Non-conformists. Non-conformists believed that people in a privileged position had a special duty, perhaps a right, to move civilization forward. In a long career of invention and discovery in the hinterland that makes him look like a character out of a novel by H. G. Wells, Cayley invented caterpillar tracks for tractors and tanks (he called it "the universal railway"), self-righting lifeboats, and an artificial hand. He drew up seat belts and automatic signals to make railway crossings safer. While working on landing gear for his aircraft, Cayley created lightweight spoked wheels that in time would lead to bicycle wheels. He

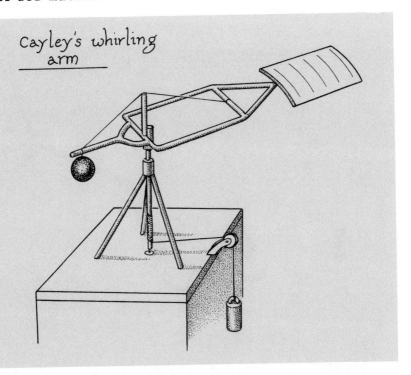

Cayley's whirling arm

also served as a member of Parliament and helped found the Polytechnic Institute in London.

Cayley was a methodical man who desired to take the measure of things, such as the number of wing strokes per second for pigeons. In March 1808 a neighbor in Brompton might have seen Squire Cayley sprinting for fifty yards in knee breeches and shirt while gripping a stopwatch. He wanted to measure the maximum velocity that an average human could contribute toward the takeoff of a man-powered aircraft.* Cayley once recorded the growth of his fingernails to determine the time to grow one full thumbnail: it was four months. Cayley was a fly fisherman and upon examining his catch one day, surmised correctly that the proportions of trout bodies would be a better shape for the fuselages of aircraft than bird bodies. From his examinations of birds Cayley drew conclusions about the best profile of an aircraft wing. He turned to the whirling-arm rig, testing a wing modeled on the crow. Later helicopter inventors

* For the baronet, that was 13.6 miles per hour.

Principal helicopter types

Single main rotor
(Bell JetRanger)

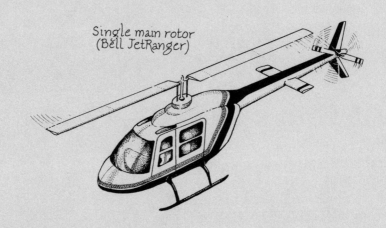

Side-mounted rotors
(Focke-Wulf Fw 61)

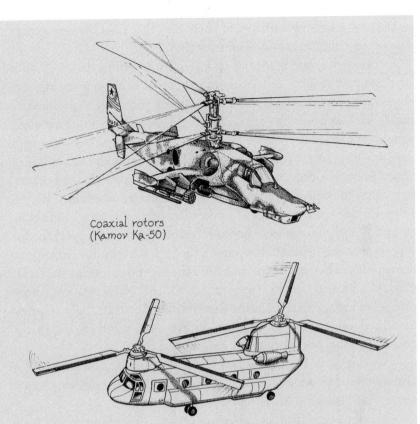

Coaxial rotors
(Kamov Ka-50)

Tandem rotors
(Boeing CH-47)

Intermeshing rotors
(Flettner Fl 282)

would use whirling-arm rigs in their day also. Today's researchers depend on computers and wind tunnels.[3]

Cayley was the first to apply a whirling-arm rig to study how air flows past objects, called aerodynamics. When using the rig, the object to be tested is fastened at the end of a pole. The experimenter stands back and sets the rig to spinning with the help of a falling weight (as Cayley did) or, for larger rigs, with waterpower or a steam engine. Using such a rig, Cayley could check for lift and air resistance under varying speeds, wing shapes, and wing angles. He used the information to improve his first scale-model glider, which he tested in 1804. Cayley realized that the undersides of glider wings should be curved slightly, in concave fashion, because bird wings worked this way. Beginning in 1809, drawing on his experiments, his correspondence with other scientists, and his exhaustive reading in the field, Cayley published a set of papers titled "On Aerial Navigation," now recognized as the first scientific description of aircraft flight. Cayley identified four distinct problems and went on to sort them out mathematically: they were lift, drag, weight, and thrust.

Cayley built two human-carrying gliders, the first ever to fly. One swooped down a hillside at Brompton Dale in 1853, carrying a servant not named in his otherwise copious notes.*

Cayley knew from his experiments that something would have to replace steam engines before aerial navigation would be practical. In 1807 Cayley experimented with an internal combustion engine that burned one pound of gunpowder fuel per minute, but mechanical problems prompted him to drop the work.

While Cayley made more progress with airplanes than he ever did with helicopters, he reached out to that field as well. In the April 8, 1843, issue of *Mechanics' Magazine,* Cayley published a detailed design for a boat-framed, beak-nosed helicopter, which he called the Aerial Carriage. He proposed one set of rotors for lift (called the elevating fliers) and a smaller set for propulsion once aloft. It would bring freedom to roam throughout the atmosphere.[4]

When Cayley died in 1857, helicopter designs were multiplying

---

* Most likely his groom, John Appleby. Cayley's achievement attracted little attention, even in England, until 1974. That year a facsimile of his man-carrying glider took to the air after being towed aloft by a car, and historians took a new interest in his notebooks and publications.

rapidly—so rapidly that the only way to make sense of how development proceeded is by grouping the helicopters in the way that biologists sort through fish and fowl. A simple taxonomy will help keep the profusion of proposals straight.

Any functional helicopter must have rotors for lift and control, some kind of engine, a mechanism for controlling altitude and direction, and a frame to hold everything together. A rotor is a set of blades mounted on a hub that spins, whether or not it produces lift. A ceiling fan is one type of rotor; a lawn mower is another. There are five ways that helicopters can combine these elements and still hop into the air. The five types of helicopters are single main rotor, coaxial, side-rotor, tandem, and intermeshing.

The single-main-rotor helicopter is the most common. News helicopters, medical helicopters, and personal helicopters are all of this type, meaning that one main rotor provides all the lift.

If there is to be more than one main lifting rotor, these must come in pairs, and furthermore each member of a pair that is driven by an engine must turn opposite to the other. Otherwise, in a demonstration of wrathful physics, the aircraft will spin out of control because every action has an equal and opposite reaction.

The subject of paired rotors leads to the second major type of true helicopter, the coaxial one. Coaxial refers to rotating items that share a common center of rotation. Take the face of a mechanical alarm clock: the minute and hour hands have coaxial spindles. To understand how the coaxial design might work on a helicopter, think of modifying a standard beach umbrella like so: slip a short metal tube over the wooden shaft of the umbrella and attach a second canopy to the upper portion of that outer tube. Now there are two concentric shafts, capable of rotating independently, each with an umbrella canopy, spaced about a foot apart. With a little gadgetry and gearing it would be possible to make the stacked, coaxial umbrellas spin in opposite directions. A helicopter with two rotors in a coaxial layout doesn't need a tail rotor, so it is more compact. This is good. But engineering is trapped in a world of trade-offs, and coaxial helicopters have other difficulties, such as a tendency for the upper rotor to throw unwanted turbulence on the lower rotor. Also, coaxial rotors, if placed too closely, can crash together.

Side-rotor helicopters rely on pairs of lifting rotors mounted on wings or side-booms. Cayley and many others assumed this was the most stable arrangement. The first successful German helicopter was of this design.

Tandems are the fourth class of helicopters. A tandem helicopter has a pair of lifting rotors spaced front to back. The CH-47 Chinook and CH-46 Sea Knight are tandem helicopters.

If the rotors are brought very close together it is an intermeshing-rotor helicopter, the fifth class of helicopter. If the two rotors are angled slightly outward, the hubs can be located just a few feet from each other without their long blades colliding. Intermeshing designs are now rare.

Using this taxonomy to classify the many helicopter designs after Sir George Cayley primed the pump, the great majority of designs that appeared were side-rotor helicopters. This seemed the stable and logical way to go. Such aircraft appear more symmetrical than single-main-rotor helicopters do, and this is perhaps why they held a strong attraction for engineers up through 1900. To novices, having a pair of lift rotors may also appear to offer redundancy, a chance for a safe landing in case one rotor breaks away, but that's an illusion. If any lifting rotor fails in flight, a multiple-rotor helicopter will go fatally out of balance unless it has so many main rotors that losing one or two won't make a difference. While Jules Verne dreamed up such a many-rotored aircraft, the *Albatross,* no helicopter has flown with so many rotors.

Despite the general agreement on the need for multiple lifting rotors, the details remained elusive. Early inventors including Paul Cornu assumed that pilots could control their ships by changing the speed of the rotors with a throttle or by using rudder-like vanes below the rotors to deflect the airflow. Neither method yielded a controllable helicopter.* Adding or cutting back on a throttle acts too slowly to offer the pilot good control, because the rotor has too much momentum. While vanes can control a hovering helicopter, they work poorly after the helicopter begins moving forward

---

* One of the last designs to use throttle control for guiding a helicopter was the Piasecki PV-2 prototype of 1943.

through the air. But there were older ideas about how to operate a helicopter with exquisite control. These concepts would come forward in due time. One of the most promising concepts came out of the windmill business.

Following the bowstring-powered helicopters popularized by Launoy and Cayley, the next set of models experimented with tipjet designs, whose rotors were powered by thrust coming out of nozzles. Horatio Phillips of England relied on combustion gases generated by burning a gunpowder-like mixture inside a chamber. Witnesses said his coaxial model of 1842 rose into the air and traveled some hundreds of feet before crashing; if so, it was the first helicopter to operate on chemical rather than stored mechanical energy.*

The potential of new fuels interested Mortimer Nelson of New York City, who won a patent for his Aerial Car in 1861 (a side-rotor design) and immediately set off in search of $10 million in capital. Nelson held in his hand the first American patent issued for a heavier-than-air flying machine.[5] He was a printer of greeting cards in lower Manhattan, located at 444 Broome Street, a few doors off Broadway. His business prospectus is of interest because he was one of the first in the field to give serious thought to the practical problems of rotary-winged flight.

Nelson planned an aircraft about thirty-three feet long, weighing a half ton. It would carry two sets of rotors, one big pair for lift and one little pair to act as propellers, all driven by a forty-horsepower steam engine. It would burn petroleum oil, naphtha, or perhaps a new type of high-power fuel he had invented, called carbo-sulp-ethal, a sort of supercharged whale oil.

Once the Aerial Car took to the air, twin propellers at the rear, near the rudder, would drive it at speeds up to 180 miles per hour. Voyage done, the craft would alight upon two metal bars that today we'd call landing skids. Nelson was the first to suggest this type of landing gear. Further, understanding that each unnecessary pound

---

* Phillips was a fan of chemistry in general and went on to invent a fire extinguisher that he called the Fire Annihilator, an iron-sheathed device that produced carbonic acid upon activation. Phillips designed it to be heaved into a burning building. Representing the American Fire Annihilator Company on the East Coast was showman P. T. Barnum, but even the bombastic Barnum lost headway in the business after a highly publicized demonstration in New York in December 1851, which went awry and allowed a two-story demonstration house to burn to the ground.

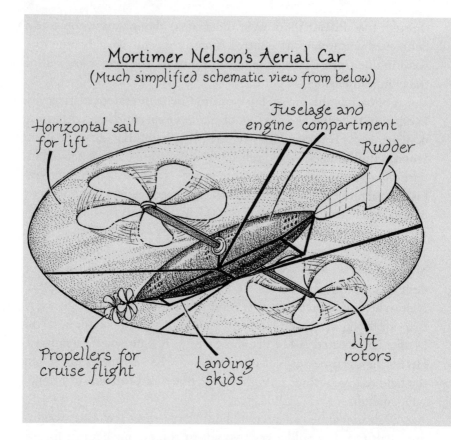

Mortimer Nelson's Aerial Car
(Much simplified schematic view from below)

Fuselage and engine compartment

Horizontal sail for lift

Rudder

Propellers for cruise flight

Landing skids

Lift rotors

must be an enemy of flight, Nelson suggested that an exotic new metal known as aluminum might work best for structural parts. While Nelson's promotional packet acknowledged than an Aerial Car was going to be expensive, whether built of wood or aluminum, in time the cost "would be accounted as of no comparative moment when weighed in the balance with a great and important enterprise, in which the nations of the earth must feel a common interest."

One of those nations was certainly France. Perhaps because of the nation's steady string of successes with ballooning, its citizens waxed optimistic about taking to the air in helicopters as well. One manifestation was Jules Verne's preachy novel of 1886, *Robur the Conqueror*.[6] The hero is a brilliant inventor-captain of the same head-strong inclination as Captain Nemo of *Twenty Thousand Leagues*. Robur flies his thirty-seven-mast, seventy-eight-rotor ship-shaped helicopter to Philadelphia and kidnaps two American men from the Weldon Institute: Uncle Prudent and Phil Smith. The Weldon Institute is a pro-balloon special-interest group. Betraying Verne's opinion on the subject, much of the book's length is spent in laying out the shortcomings of balloons. Robur expects that the voyage will convince Prudent and Phil that powered balloons can in no way measure up. Instead, Robur's obstreperousness convinces them to plan on blowing the helicopter up with two pounds of dynamite.

The rotors are powered by secret batteries of such staying power that the ship never pauses to recharge. The craft is named *Albatross*. While such a name would be an unfortunate choice for ships or aircraft in our time, in that era it signified a noble bird that is thoroughly at home on the ocean swells. Before the detainees get around to touching off their charge, the *Albatross* hovers at low altitude to tow a boatful of shipwreck survivors to safety. Verne was the first writer to depict a rescue by helicopter.

"With her, I am master of the seventh part of the world," Robur informs his passengers. That seventh part is "larger than Africa, Oceania, Asia, America, and Europe, this aerial Icarian sea, which millions of Icarians will one day people."

Verne's interest in helicopters dated to 1863, the founding of the "Society for Encouraging Aviation Using a Machine Heavier Than Air." A handful of French citizens set up the Heavier Than Air Society and helicopter-history books usually emphasize the role of the mechanic named Ponton D'Amecourt. D'Amecourt built a series of scale models for the society, some powered by springs and some by steam.

D'Amecourt notwithstanding, the real force behind the society was an extraordinary artist, extrovert, chaser of fame, and lover of gadgets known to cultured Parisians by his boyhood nickname of Nadar. Like Nelson, Nadar was a helicoptrian. Verne admired Nadar greatly. In his novel *From the Earth to the Moon*, Verne based the character Michael Ardan on Nadar, as is evidenced by the anagram.

Born to a publisher and his wife in 1820 as Gaspard Félix Tournachon, Nadar dabbled in medicine as a young man but turned to satirical writing and caricature, wielding a wit that was cutting enough to catch the attention of the Bohemian culture then emerging in Paris. At first he planned to draw a thousand portraits, gathering up every person of significance in the Paris social scene, but gave it up before finishing in favor of portrait photography. It was his rare skill at capturing personalities in photographs—which are still of interest to art historians today—that brought him commercial as well as critical success beginning in 1854. Encouraged by the public reaction, Nadar set up a photographic studio two years later that attracted many celebrities. But he was prone to ennui, and to seek out new vistas he went out to photograph the Paris sewers. In doing so he was the first photographer to employ artificial lights, which in his case were magnesium flares. Nadar's collection of sewer photos did not bring commercial success, but by then a new interest was upon him.

Nadar was the first person in history to take photographs from the air, by going up in a balloon tethered over the Bievre Valley in 1858. This experience with a balloon caught his fancy and soon he was spending more time around balloons than at his photography studio. His finances would never recover, but the flair he brought to flying balloons would bring him the international attention that his photography had not.

Nadar maintained a shop in Paris for assembling these balloons and it was here in September 1863 that he convened a meeting of friends and acquaintances who might support him in taking a new direction in aviation. The meeting meshed well with the fast-moving times: in the United States, the Transcontinental Railroad was getting under way, and London had begun service on the first section of subway system. Jules Verne had just completed *Five Weeks in a Balloon*.

Nadar explained at the meeting that for all the joys of ballooning, the avocation frustrated him since he was at the mercy of the winds. "To conquer the air," he announced, "one must be heavier than air." Accordingly, he invited them to join in the Heavier Than Air Society. Gabriel de la Landelle was on hand to set one of the D'Amecourt helicopter models going. The time was right, Nadar said, to build a full-scale D'Amecourt aircraft, an "aerial locomotive" good for making world-girdling voyages. He would lead the way in raising money.

To that end, Nadar said, he was going to fly an enormous bal-

loon, called *Le Géant*. Nadar promised it would bear eighty people
aloft (plus a printing press) for eight days and eight nights. He said it
would lift off from the racetrack at Baden-Baden, Austria, proceed to
London, and thence to the Mediterranean.

Surprisingly, given his shaky finances, Nadar went on to build
Le Géant and started inflating it a month later to raise money for his
helicopter. As built, it was a double balloon carrying a gondola,
standing 196 feet high and using 22,000 square yards of silk. The
gondola, made of rattan and ash wood and given shape by inflated
rubber tubes, was two stories tall. It had a captain's cabin, four passenger cabins with bunks, a liquor cabinet, a photography studio, a
washroom: all this packed into just 360 square feet of floor space.
One experiment showed that thirty-five soldiers could fit onboard.
Passengers reported the impression of being confined in an oversized
picnic basket.

Parisians by the tens of thousands gathered to see the first flight,
originating in the Champs de Mars in Paris, on October 4, 1863. The
trip ended far short of Nadar's projections and the gondola flopped
onto its side. But an even larger crowd came to watch the second
launch. *Le Géant* crossed Belgium and the Netherlands. It came down
with a thud in Nienburg, Germany, with the gondola once again
touching down at high speed but this time tipping over on a railroad
line. The voyage looked to be on track for a spectacular and memorable end, because a train was approaching. As Nadar recounted later,
unapologetically, "A single cry escapes our throats, but what a cry!"
Passengers erupted out the windows as the envelope dragged the
gondola toward what was looking to be the world's first balloon-
train collision. An assistant hacked at the balloon shrouds with an ax
and it tangled to a halt in a telegraph line. The locomotive stopped
short, the envelope was left mostly intact, and no one was fatally en-
trained, but the experience did much to chill interest in more flights.
As a British magazine said, *Le Géant* had "all but carried all the passengers to death's door." Nadar continued with smaller balloons and
was back in the news during the siege of Paris in 1870. According to
French legend, Nadar fought a high-altitude gun duel over the outskirts of Paris with a Prussian, each in his respective balloon.

In the end Nadar did not build an Aerial Locomotive, any more
than Nelson constructed an Aerial Car; the dream took flight only in

Verne's novel. Even D'Amecourt's elegant, aluminum steam-driven scale model didn't lift off the ground while witnesses were watching. But one lasting thing arose out of the great enterprise: Ponton D'Amecourt's new word for the device. Henceforth, with a literary boost from Jules Verne, the aircraft would be called the "helicoptere," meaning helix-wing.

Cayley, Verne, and the early balloonists popularized the notion that while the atmosphere could be seen as an ocean, it was going to be a better place for travel than any saltwater sea. The domain of Icarus lapped at everyone's door, which meant a whole new way of moving about. "I think it a national disgrace in these enlightened locomotive times not to [raise money] which would secure to this country the glory of being the first to establish the dry navigation of the universal ocean of the terrestrial atmosphere," Cayley wrote in 1843. This vision was compelling to his contemporaries because ships could cross whole oceans in less time than it took travelers to go a few hundred miles down rugged roads.

These aircraft would have to be powered. Gliders had no range—Cayley's had flown less than 500 feet at Brompton Dale—and early balloons did no more than float with the wind. If mankind's strivers found nothing better than steam for propulsion, they would do little better than float dreamily like the characters in Eugene Field's poem "Winken, Blinken, and Nod," adrift in their wooden shoe, in a sea of air.

Chapter Three
A New Natural Power

By the mid-nineteenth century, balloon builders found it much easier to get moving than to arrive at a particular destination. Manned balloons, and in fact all kinds of aircraft under consideration at the time, needed a breakthrough in propulsion. One proof of this proposition was a test flight in 1859 made by the American balloonist John Wise.

Wise had once tried to persuade Congress to finance a bomber-balloon for the Mexican-American War. He had promised to use it to drop explosives onto the walled fortress at Vera Cruz. Congress didn't finance the plan, but Wise stayed in the business. From his later high-altitude measurements he realized that were a balloon able to climb above 12,000 feet and stay there, it could cross the Atlantic by using the predictable mass of air that sweeps from west to east. He prepared a test for July 1859: He would transport a bag of mail from St. Louis to the East Coast in his balloon *Atlantic,* and set a new distance record. Three passengers went along: his financial backer, Vermont pottery manufacturer O. A. Gager;

balloon engineer John LaMountain; and a St. Louis newspaper reporter. These three rode in a wooden lifeboat. Wise perched above
them in a wicker basket.

Atlantic did well until it met a storm over Lake Erie. It dropped
to within a few feet of the waves and accelerated under the straight-
line winds. Dumping out everything, including the U.S. mail, the
passengers of the *Atlantic* reached shore before it crashed into an elm
tree, tearing off a hefty limb. The *Atlantic* had covered over 800 miles
at an average speed of forty miles per hour. It was, said Wise, "the
greatest balloon voyage that was ever made."

The solution to aerial drift seemed to be engine-powered balloons. The first to drive a balloon of any kind under power was
Rufus Porter. In 1847 Porter attached a clockwork spring to a small
dirigible and demonstrated it in New York City. After the discovery
of gold at Sutter's Mill, he built more models, gathered investors, and
actually started to build a hydrogen-filled airship to be powered by a
steam engine. This he expected to haul four dozen passengers at one
hundred miles per hour. As part of the promotional effort, Porter advertised for advance ticket sales and published a book titled *Aerial
Navigation: The Practicality of Traveling Pleasantly and Safely from New
York to California in Three Days*. Porter never got to finish his behe-
moth but went on to found *Scientific American*.

The era of purposeful travel by air really began on September
24, 1852, when Henri Giffard, a French engineer, rose above the
Hippodrome in Paris on a cigar-shaped blimp filled with flammable
gas. Driving two propellers, Giffard's coal-fired steam engine was
powerful enough to move the blimp at six miles an hour. Contemporaries saluted him as "the Fulton of aerial navigation."

Most steam engines carried on model aircraft of the nineteenth
century relied on very hot water stored in a tank at high pressure.
One of the best performers was a fixed-wing model airplane using a
highly advanced steam engine created by F. J. Stringfellow for a competition held at the Crystal Palace Exhibition in 1868. Stringfellow's
engine produced a full horsepower* and its reservoir of pressurized
water was enough for ten minutes of operation. When demonstrated
to the public, the airplane flew along a guide wire for safety, but the

---

* One-horsepower industrial steam engines of the period weighed a full ton, so a sixteen-pound steam
engine with the same output was remarkable.

engine was powerful enough to sustain the aircraft in level flight during one unofficial test in the thousand-foot-long building.

An editorial in an 1860 issue of *Scientific American* predicted that aircraft should be able to travel at the rate of a mile per minute. "What a luxurious mode of locomotion! to sweep along smoothly, gracefully and swiftly over the treetops, changing the course at pleasure, and alighting at will; how perfectly it would eclipse all other means of travel by land and sea! This magnificent problem, so alluring to the imagination and of the highest practical convenience and value, has been left heretofore to the dreams of a few visionaries and the feeble efforts of a few clumsy inventors.... The thing that is really wanted is a machine driven by some natural power, so that the flyer may ride at his ease. For this purpose, we must have a new gas, electric or chemical engine."

There were more dead ends than live ones in this search. "The crop of credulous capitalists never grows less," reported the magazine *Forum* in 1888. "Ether-engines, bisulphide-of-carbon engines, ammonia-engines, and carbonic-acid-gas engines, cloud-engines, and chloroform-engines come and go with all the certainty, if not the regularity, of the seasons, and each lives its short life and disappears, only to be succeeded by another of the same tribe."

The dream refused to die, even so. Airships with some radically improved form of propulsion would be ideal for probing the world's nether regions, particularly those areas unfriendly to ships. By the 1860s explorers had reason to pine for alternatives to ships, particularly in the hull-snapping pack ice of the Arctic. One example was the voyage of the wooden-hulled ships *Erebus* and *Terror,* which Sir John Franklin had hired in 1845 to map the Northwest Passage. After those ships and Sir John were declared missing in 1848, a decade of expeditions followed to search for survivors, bodies, or relics.

Erebus and *Terror* had twenty-horsepower steam engines, which are classed as external combustion, meaning that the source of heat comes from outside the section that produces power. An example of such a heat source is the wood-stoked firebox on an old-fashioned locomotive. Early engineers knew that steam engines were profligate with water and fuel. They turned to the contemplation of "hot-air" engines, where the pistons would be driven by exhaust gases from separate combustion chambers.

Other engineers speculated that some kind of explosive compound, ignited in the piston itself, would perform even better. These would be internal combustion engines, meaning that the useful heat would be generated inside the engine itself.

The origin of internal combustion engines goes back to 1680 and the Dutch physicist Christian Huygens, who drew up plans for burning gunpowder inside a cylinder to drive an engine. This was before steam had proved itself. Huygens's protégé Denis Papin looked the idea over but dropped it because the explosions left residues that spoiled the cycle. Soon Thomas Newcomen proved steam engines more workable and these were a roaring success by 1707. All such engines were huge and heavy, but that was not a problem given that they were anchored to the ground. Pumping water out of deep British mines was the original function of the engines, and the pumps were so useful that the customers were happy to pay for steady improvements. By 1800 the biggest steam engines were approaching 200 horsepower.

Still the hot-air engine beckoned as a potential paragon of thermodynamic efficiency, light in weight and high in power. When Thomas Edison began testing helicopter rotors, he experimented with an aircraft engine powered by a ticker-tape made of explosive guncotton. Spooling off a supply roll, the tape served as fuel when pulled into an engine. After the feed-roll detonated during tests and burned his assistant, Edison dropped the idea.

Flammable liquids and gases were more promising as fuel for such engines. Jean Joseph Étienne Lenoir built the first commercially successful hot-air engine, one that burned gas derived from coal. Lenoir's early coal-gas engines found immediate users in urban factories and shops that needed compact powerplants to drive the belts for their machine tools and presses. Lenoir's two-cycle engine had spark ignition, and later a carburetor to mix air with the fuel. But it was low in power and the task was left to Alphonse Beau de Rochas and then German inventor Nicolaus Otto to think of a radically improved design, one that compressed the fuel-air mixture before igniting it. This raised the horsepower produced per pound of engine. Powerplants for the Wright Flyer and for many other aerial firsts grew out of work by Otto and de Rochas.

Lenoir deserves credit for the way his engine business inspired

Otto to leave his job as a salesman in the sugar-and-flour trade and take up motors. Lenoir should also get a nod for his experiments with "rock oil." He was looking for something more convenient than coal gas and cheaper than whale oil or kerosene fuel. In his day, kerosene was a lamp fuel distilled from coal or the tarry goo called asphaltum. Lenoir's second generation of engines relied on the new fuel.

Today, when oil is pumped from rock we call it crude oil. Edwin Drake had drilled the first successful oil well only a few years earlier, in 1859. A corps of eager drillers spread worldwide and developed many fields, including ones in Venezuela and Eastern Europe, but by the end of the century there was reason for frustration about oil from rock.

It was the price. Refined petroleum had established itself as the most desirable fuel for powering vehicles, but it loomed as prohibitively expensive for use in millions of engines. Most American crude oil came from a heavily exploited zone stretching from Ohio to New York, and there was a promising flow from Russia's new Baku field. But even taken together these produced more of a stream than a torrent, and in the United States the Rockefeller interests controlled so much production they set the price nationwide.[1] Through 1900 there was a distinct possibility that only the wealthy few would ever be able to use gasoline engines in their boats, dirigibles, and motorcars.

The first indication that petroleum would be plentiful enough to run a new century came just days after that turn of the calendar. In January 1901 a crew called Hamill Brothers was using a steam engine and an early version of a rotary bit to bore into a low hill on the coastal plains of Texas, near Beaumont. Earlier, the Hamills had encountered soft ground that threatened to cave in and choke the hole. They invented drilling mud and kept on going. By 8:00 A.M. on January 10, the bit had reached a thousand feet of depth and the crew started to pull the drill string to replace the bit. At this moment the man in charge of the project, mining engineer Anthony Lucas, was in a general store in Beaumont. Lucas had been so optimistic about the signs of oil around there that he had been selling furniture out of his house in town to pay living expenses. It did not discourage Lucas that everyone said all the real oilfields lay far to the northeast.

As the Hamill crew hauled up the iron pipe, the men felt a rumble. Mud and then oil rose up the bore with such force it carried iron pipe hundreds of feet in the air. The stream of warm green oil was six inches across. Quickly the geyser reached a height of 200 feet and kept getting bigger, as if some gargantuan pipeline had been breached. Nothing like this had ever happened in Pennsylvania, even when those fields were fresh and full. Lucas arrived out of breath—on foot, because a carriage wasn't fast enough to suit him—and hired crews to throw up earthen dikes to save the oil for sale. Soon an artificial lake fifty acres in area wasn't big enough. That which overtopped the dikes gathered itself into a creek of crude oil that in time reached the Gulf of Mexico, thirty-five miles away. "The volume of oil if anything seems to be on the increase," a Galveston bank official wrote home on the fifth day.

It took nine days to get the well fully under control. For a time this one well yielded twice as much oil as the entire state of Pennsylvania. The most productive patch of the Spindletop Field was only fifteen acres in size—the area of a shopping mall's parking lot today—but it turned out more oil in its heyday than the rest of the world. While that pace couldn't be maintained for long, it lasted long enough to set off a frenzy of oil exploration among other Gulf Coast salt domes. The price of oil in Texas fell from a dollar per barrel to twenty-five cents, and then to two cents, hitting the lowest price in history. Spindletop broke the grip that Rockefeller interests had held on American petroleum. After inspecting Spindletop, the mayor of Toledo, Ohio, concluded, "Liquid fuel is to be the fuel of the twentieth century. Smoke, cinders, ashes, and soot will disappear along with war and other evidences of barbarism."

One aviator of the new century who depended greatly on the rapidly improving fuels and engines was Alberto Santos-Dumont. Santos-Dumont popularized a new way of thinking about aircraft travel.

Before Santos-Dumont's fame, the generally accepted goal of air travel was the grand voyage, whether by balloon (Verne's *Five Weeks in a Balloon*) or helicopter (*Robur the Conqueror*). Each ascension of *Le Géant* in 1863 was an adventure to be celebrated. As the Princess la Tour d'Avergne boarded Nadar's big balloon for a trip in 1863, she sent a note to her friends with these instructions: "Do not sit up for me. I shall not return tonight. Perhaps not tomorrow night. Perhaps never." Each balloon trip out of Paris during the siege of 1870 was likewise an adventure.

Santos-Dumont trimmed this glorious vision down to something useful by building aircraft sized for daily errands rather than epics. Santos-Dumont was born in Brazil to a wealthy family of coffee growers and showed an early interest in machines. He relocated to Paris at age eighteen to study chemistry, physics, and engineering. His father's death left him a fortune and he was able to pursue his childhood interest in ballooning. On October 19, 1901, he won the "Tour Eiffel" Prize by making a seven-mile flight with a powered dirigible in less than thirty minutes, under conditions that required a U-turn around the Eiffel Tower.

One result of the jaunt was a craze for men's wristwatches, which had never been popular before. After Santos-Dumont told jeweler Louis Cartier that he had had no time to look at his pocket watch during the trip, Cartier produced a model named after the short, dapper Brazilian and sold it around the world.

Santos-Dumont went to work on a smaller model, *Santos-Dumont No. 9*, nicknamed the *Stroller*. Less than fifty feet long, it was powered by a three-horsepower engine and no bigger than needed to carry the short, lightly built Brazilian in a wooden framework slung below. He used the *Stroller* at every opportunity, commuting with it between Neuilly St. James and St. Cloud. When dropping in on friends in Paris, he tied the *Stroller* to their balconies, leaving it to loom over the sidewalk, nestled between buildings on one side and the boulevard trees on the other.

For all his love of powered balloons, Santos-Dumont made initial plans to build a tandem helicopter, before deciding that gasoline engines could not sustain them in the air. But experimental airplanes were ready for adventure, so like dozens of others, he turned to that field, leaving others to wrestle with the unruly helicopter.

Chapter Four

Breakout

In early January 1907, a New York State manufacturer by the name of Glenn H. Curtiss was preparing for the Florida Speed Carnival. The race was an annual contest between one- and two-cylinder motorcycles, to be held on January 23 on the hard-packed sands of Ormond Beach, north of Daytona Beach and fronting the Ormond Hotel. Completed in 1890, the Ormond Hotel still offered a glittering retreat for northerners this late in the Gilded Age.

For entrepreneurs and dreamers alike it was an immensely exciting time to be alive; suddenly the open spaces were opening up to travel and development. Blimps showed some promise, but whatever aircraft accomplished, motor vehicles and roads looked likely to break the long monopoly of railroad barons over long-distance transportation, comparable to how Spindletop had broken the back of oil monopolists. Speed was only part of this revolution, but what a part: Vehicles competing in 1907 were expected to hit at least 120 miles per hour on the hard wet sand of Ormond Beach. Only six years earlier

THE GOD MACHINE   51

members of the American Automobile Club had been cited for driving thirty miles per hour on a road marked for ten miles an hour.

It was a time of "speed intoxication," said Russian mathematician George de Bothezat the following decade. "Speed of transportation means also the increase of human lifetime, which always was and is one of the most burning desires of man," de Bothezat wrote in an article, predicting that airplanes crossing the oceans would bring a new era of world harmony. "[If] the course of all humanity is limited, let us at least make the best use of the time left to us. Speed is our most powerful ally. Let us use it, develop it, and venerate it. It is in infatuation with it that is found the most powerful source of happiness."

The greed for speed created an unquenchable demand for high-performance gasoline engines, a development that seemed to promise an early breakthrough for manned helicopters. But helicopters' first lumbering hops would reveal serious problems of control and stability. Engine power, it turned out, was a necessary but by no means sufficient condition for successful helicopter flight.

Glenn Curtiss made ready a stable of his Curtiss-brand motorcycles to enter in conventional categories at the Speed Carnival. He decided to go the extra mile and try a custom-built bike mounting an air-cooled V-8 engine that he had designed for aircraft. The engine would be four times bigger than other powerplants, so powerful that no class existed for it. It would be a thrill even for the steel-nerved Curtiss, who in setting multiple world records rode with such abandon that he was already known to the public as "Hellrider." In later years Curtiss and his town of Hammondsport would be the inspiration for the first Tom Swift books.

Flamboyant titles may have seemed out of place for such a shy person, but it was based in fact. Curtiss was the first American motorcycle champion, he built and sold the country's first private airplane, and he made the first official flight in the country in 1908. He also earned the first U.S. pilot's license; his engine powered the dirigible *California Arrow* that made the first round-trip journey in America, in 1904; and he built the navy's first airplane. In 1919 his NC-4 flying boat was the first aircraft to cross the Atlantic.

He lacked a suitable motorcycle chassis—no frame and transmission of the day could handle such a monster engine—so he directed his mechanics to build one. Time was so short that his men loaded the V-8-powered bike aboard the railcar for Florida without ever having road-tested the vehicle. All Curtiss really knew, from coasting down a hill on it one time, was that he could probably balance on it well enough to ride. Airship pilot Tom Baldwin came along to see what the motor would do when opened up.

After the sanctioned Speed Carnival races were over, it was time for speed, pure and simple, over the measured mile. A Stanley Steamer automobile labeled the *Rocket Racer* made its first run, but failed to beat its previous years' record of a mile in 28.5 seconds. On January 25, driver Fred Marriott turned up the steam pressure for a second try but hit a small rise on the beach. His car flew into the air at approximately 120 miles per hour and broke apart on landing. Marriott was seriously injured.

Curtiss was unruffled by the condition of the beach when he prepared to make his run, though he still hadn't tested the vehicle. After a push-start from his crew, he rolled open the throttle on the handlebar—the handlebar throttle was another Curtiss invention— and headed for the starting line down the beach. Somehow he kept control despite the beach's irregularities. After he passed the spectators who were assembled at the finish line, it took him another mile to stop.

Officials at the Speed Carnival clocked Glenn Curtiss at 136.4 miles per hour. No machine of any kind had gone this fast before. It would be another four years before any manned vehicle would beat Curtiss's record.

The Ormond Beach trial wrecked the transmission on Curtiss's V-8 motorcycle. It never ran again. One of the consequences of the Ormond Beach speed record was a note from Alexander Graham Bell asking Curtiss to pay him a visit. This refreshed a connection that Bell and Curtiss had made in January 1906 at the New York Auto Show.[1]

Bell's invitation following the Ormond Beach run pulled Curtiss into the Aerial Experiment Association. By mid-1908 Curtiss was well on the way to his historic role as aircraft developer, which ultimately would prove much more important than his work on motorcycles.

"The sons of Daedalus are indeed a rapidly increasing tribe," wrote Harold Howland in a 1907 issue of *Outlook* magazine. Airplanes were appearing on both sides of the Atlantic, airships could be rented for private outings in Paris, and news reports from France assured readers that helicopters were finally taking to the air.

From 1900 through 1930 many rotary-wing developers laid claim to having built the "Wright Flyer of helicopters," meaning the one that made the first real flight with a real person on board. But any single claim can usually be disputed with a counterexample or, in some cases, by a skeptical look at the documentation. Reasonable minds could also differ as to which rotary-winged performance would count as truly equivalent to the Wrights' straight-line flights among the Kill Devil Hills of North Carolina's shore. The threshold could be hovering at low altitude without assistance from the ground, or traveling one kilometer without crashing, or making a round trip at a useful cruising speed.

For years helicopter histories have listed the first helicopter to rise and hover as a simple machine built by Jacques and Louis-Charles Bréguet[2] in 1907, or alternatively one that Paul Cornu tested later that year. Emile Berliner and his partner, John N. Williams, might have achieved a hovering flight at a Maryland farm in June 1909, during the first of four great waves of helicopter progress.* Or was the first hover achieved by Jens Ellehammer's Danish-built coaxial helicopter, which hopped into the air in 1912?

While helicopter historians generally agree that no helicopter achieved true flight until the early 1920s, the identity of that first hoverer still makes for a good mystery.

Paul Cornu, born in 1881 in Lisieux, France, was inspired into aviation in 1906 by a cash prize offered to whoever made the first round-trip flight of one kilometer. He gathered money from over a hundred friends and built a tandem helicopter. There were two rotors mounted atop a four-wheeled cart, driven by a twenty-four-horsepower engine. Not coincidentally, for Cornu was a bicycle

* The first wave happened from 1907 through 1912, which produced helicopters that hopped briefly into the air; the second was in the early 1920s, and produced helicopters that could hover and move short distances; the third began in the late 1930s and produced helicopters capable of travel and lifting loads; and the fourth was the arrival of gas-turbine-powered helicopters in the 1950s.

maker, each rotor was built around a belt-driven pulley that looked much like a big bicycle wheel. Cornu bolted two short, silk-covered rotor blades to each pulley's rim for lift. Its lightweight frame and these oval, translucent wings, and two more such panels mounted below for directional control, gave Cornu's machine a dreamy appearance, like a limousine for elves. Cornu didn't try for the Aviation Grand Prix with it, but he did claim in an article published in 1908 to have fluttered from the ground at Lisieux on November 13, 1907, with himself as pilot. He further claimed to have repeated the feat with a passenger, his brother, onboard.

A basic doctrine of aerodynamics called momentum theory* defines the outer limits of how much lift a specific helicopter rotor can provide. It refutes the claim, according to Professor Gordon Leishman of the University of Maryland's aerospace program. Whatever the merits of the lifting surfaces, a twenty-four-horsepower engine could not have done more than lighten the machine on its wheels and, at most, would have enabled the machine to hop in the air briefly if a gusty wind had assisted by swirling under the rotors. Under ideal conditions, Leishman says, six more horsepower would have been required to raise the helicopter fully off the ground. In any case, no photographs have materialized that show the machine in flight. In Leishman's opinion Cornu should be credited with a serious attempt to hover but he clearly stretched the truth about his machine.

The other contender of 1907 starts with more credibility, in part because a key participant, Louis-Charles Bréguet of the famous French clock-making family, went on to a highly successful career designing airplanes and, in 1935, participated in building a fully functional helicopter prototype.

Bréguet claimed that on August 24, 1907, his Richet-Bréguet "Gyroplane No. 1" rose into the air on its own at Douai, France. It was a vast and spindly structure driven by a forty-five-horsepower engine. Bréguet said that four assistants on the ground had to hold it steady because it lacked piloting controls. But two of the assistants admitted later they had done more than that: They had helped lift No. 1 off the ground. Historians disqualify aspirants for that kind of thing.

* Momentum theory came from the work of W. J. M. Rankine, Alfred Greenhill, R. E. Froude, and William Froude.

The problematic claims of Cornu and Bréguet leave open the door for other, later contenders to seize the title of First Hovering Helicopter, or perhaps First Hopping Helicopter. Successful or not, their spindly machines received little attention at the time because any announcements came in the shadow of publicity generated by Cornu and Bréguet and therefore looked like also-rans. One such hopeful was an aging inventor who hoped to apply traditional Yankee tinkering to the new world of vertical flight, and he had the perspicacity to team up with some formidable inventors. He knew exactly where to start: Hammondsport, in upstate New York.

The center of activity for American aircraft development, covering airships, airplanes, and helicopters, was Hammondsport in the Finger Lakes region. During 1908 hundreds of aircraft enthusiasts and job seekers headed for this grape-growing town at the south end of Lake Keuka, reachable via a branch railroad line off the main track. The gathering spot for builders and onlookers was a winemaking area south of town called Pleasant Valley. The hills nearby gave it the look of a "Baby Switzerland," according to some.

The resort town of Hammondsport possessed the attractive power that the Yukon had recently held for gold-seekers, and that Silicon Valley would in time have for computer geeks and venture capitalists. The principal reason was Hammondsport resident Glenn Curtiss and his "shop," which was Curtiss's laconic term for the G. H. Curtiss Manufacturing Company. Though extremely busy, Curtiss didn't wall himself off from the throng of dreamers and tinkerers as the Wright brothers had done.

"In no other locality on earth is there such an assemblage of genius along the line of aerial work as Hammondsport can now boast of," said Alexander Graham Bell, "being attracted hither by the Curtiss motors, acknowledged to be the lightest and strongest motors made." On any given day, visitors disembarking at the train station could be dreamers like the impoverished shoemaker from Chicago who wanted to build a man-powered flying ship, all the way to notables like Henry Ford. Many new visitors arrived for the July 4, 1908, attempt to win the Scientific American Trophy. On that day Curtiss lifted off from a horse track at Stony Brook Farm to make the first officially witnessed flight of an airplane in the Western Hemisphere.*

One of the oldest of the hopefuls in Hammondsport at this time was a white-haired engineer-entrepreneur named John Newton Williams. Williams had hauled a pile of helicopter parts and sketches with him from Derby, Connecticut. The reason was that Glenn Curtiss had offered to loan Williams a V-8 engine of the same kind that made the famous motorcycle run at Ormond Beach.

A review of Williams's life up to his helicopter career shows much inventing and a wide range of businesses, but none of it had to do with aircraft and the dynamics of large rotating weights. That he would undertake such a venture at age sixty-eight without any aviation background reflects high optimism. Williams, born in Brooklyn, was just five years younger than Samuel Clemens. By 1908, with his bushy mustache and proud chin, he also looked like the old riverboat pilot and author.

In 1862 Williams had left his farming family and joined the Union Army in Minnesota, where he fought in several Indian battles. After the war he was a horse racer but shifted to the design and sale of farming and office equipment, often getting into the protracted legal battles so typical of the brutal business environment of the Gilded Age. His Williams Visible Typewriter was a competitive machine but didn't reach volume production until 1892, which was too late to join fully in the office-machine sales boom. Though he founded a typewriter factory in Derby, Connecticut, the business went into decline after 1905, and Williams handed it off to a partner. At that time he decided to build himself a helicopter.

The concept he hauled to Hammondsport was a simple coaxial design. It was little more than a square wooden platform with a big engine and a transmission, vertically connected to two coaxial rotors on a spindle. The pilot had a sliding chair to sit on, which shifted the center of gravity and therefore counted as one of the few controls. The rotor blades were tilted panels.

Williams's attempts to lift off at Hammondsport in the summer of 1908 must have left him unsatisfied, because he sought out a bigger

---

* The Wright brothers declined the organizers' offer to make a witnessed flight for the prize, though they would have won it before Curtiss.

powerplant. A chronic power shortage was the bane of all early heli-copter inventors. In April 1909, Williams relocated to Washington, D.C., the home of Emile Berliner. Berliner, a German immigrant who first made his mark in acoustics, was best known to Americans as inventor of the disk phonograph that replaced the Edison record-ing cylinder. He also invented an improved microphone that Alexander Graham Bell used for his telephones and—relevant here—one of the first rotary engines specifically for aircraft.

Rotary engines are simple and powerful (although oily), and they offer good air cooling, because the engine and attached pro-peller turn while the crankshaft remains still. Berliner was fascinated with airplanes and helicopters (which he called aerobiles). News of his work appeared often in the *Washington Post*. Berliner offered Williams the use of two thirty-six-horsepower rotary engines of his design, the Adams-Farwell airplane engine.[3]

The *New York Times* reported that, using these motors, the Berliner-Williams helicopter lifted off the ground in late June 1909 with Williams onboard. But no photograph has turned up and a press account is not conclusive.*

Still, from the technical details that Williams gave out, it would have been possible for a machine of that weight and description to have lifted a few feet off the ground with the twin engines and hov-ered there, if assistants were on hand to steady it. "It would seem that hovering would have been possible with this rig, but it was mar-ginal," according to Gordon Leishman, professor of aerospace engi-neering at the University of Maryland. "The same calculation for Cornu shows that hovering with his rig was clearly impossible."

All early helicopter pioneers must have been dismayed at the scary behavior their machines exhibited as they approached liftoff, and particularly if they rose free of the ground. It's a combination of twisting and vibration resulting from the fact that much weight is rotating at high speed and has a great deal of leverage. Williams's machine, for example, was afflicted with a rotor driveshaft that flopped from side to side. A helicopter's instability, twisting, and

---

* Press accounts of the early aviation age frequently exaggerated accomplishments, even when reporters were present at the event. An early example was coverage of Jacob Degen's "manned flight" in Vienna in 1809. Reports omitted the fact that Degen was suspended from a hydrogen-filled balloon while trying to move with wings on his arms.

vibration was much more troublesome than on early airplanes: since a helicopter is not rolling down a runway, there is no powerful flow of air to dampen these problems. A vibration can start in a shaft at one end, spread through the machine, and match a weakness at the other end that causes it to amplify quickly.

The June 1909 tests halted after an accident gashed an employee of the Berliner research laboratory. Concluding that the problems of helicopter flight were not remediable in the short term, Emile Berliner turned to other matters and Williams returned to manufacturing, this time making motorcycles and three-wheel delivery carts under the name of the New York Rotary Motor Company.[4] World War I, and in particular the U-boat war in the Atlantic, prompted him to turn again to helicopters. Throughout 1917 Williams hunted for investor money, made more tests, and tried to persuade the navy to develop the Williams-Berliner helicopter for hunting submarines from ships. Even after the navy's construction bureau wrote him that the machine was too underpowered to be of use at sea, a small group of family and friends chipped in money to keep the proposition alive for the remainder of the war.* After the war, Williams continued to tinker with the helicopter in the grassy fields of eastern Connecticut, perhaps thinking of winning a British prize for the first helicopter to reach 2,000 feet and fly at sixty miles per hour, and to land safely after an engine failure. He gained no further ground before going into a retirement home. The typewriter-helicoptrian died three years later at the age of eighty-eight.

But no one else was winning helicopter prizes either, and aviation experts didn't know what to make of this troublesome aircraft, which promised much and delivered nothing. One such skeptic was the outspoken, delightfully obsessive John Frederick Thomas Jane. Jane's series of encyclopedias on war and transportation equipment** may go on as long as civilization itself, but at the beginning he had only one obsession to his name and that was the warship.

Jane, son of a vicar in Portsmouth, England, was twenty-four years old when he left home to pursue his dream of drawing ships

---

* One investor in Williams's helicopter company was the famous early balloonist Roy Knabenshue.

** One is *Jane's All the World's Fighting Ships.*

with such detail so as to distinguish nearly identical ships within a given class—for example, Germany's *Victoria Luise* class of cruisers. This required a rare eye for detail, but even so he was unable to land his dream job in London. Money exhausted, he moved to an ocean-side embankment overlooking his beloved ships at anchor. He begged for food money. Finally Jane's portfolio of lovingly accurate ships won him a magazine assignment to render the German fleet then at anchor in Spithead Harbor.

Jane graduated from this work to publish the first edition of his encyclopedia of ships, which proved so essential that no expert dared be without a copy. His career under way, he branched into aircraft in 1909. Because Jane's first love was ships, it is perhaps not surprising that his early books on aircraft had a skeptical tone. When Jane put together the second edition of *All the World's Airships,* dated 1913, entirely absent from the main pages were the helicopters that had appeared in the first edition four years before, suggesting to the reader that this entire line of development had withered away. The few helicopters that did appear in its pages were confined in an appendix titled "Historical Aircraft," which Jane called his "cemetery of dead ideals." Of course, Jane had doubts about airplanes too, which he said were languishing as a means of private transportation: "Except as a war-machine," he wrote, "the aeroplane is of little interest or use to anyone."

That acerbic prediction missed the target, but Jane correctly forecast that military needs would drive the aircraft industry in a way that civilian purchases never could. That principle would be true of the first military rotorcraft, built by a group of Austrians and Hungarians who, halfway through World War I, began working on a replacement for the tethered "kite" balloons used by battlefield observers. Observers gathered information about enemy troop locations for use by artillery batteries and infantry commanders.

Balloons had gone to war for the first time in June 1794, when the French hydrogen gasbag *L'Entreprenant* raised an observer to peer into Austrian lines at Maubeuge, but by World War I the list of balloon vices had outgrown the roll of virtues.[5] Logistics was a problem. Each German kite balloon of World War I required 140 men and six vehicles, and the hydrogen lifting gas was dangerous to handle.

Another problem was enemy fire. By 1917 balloons were slumping to the ground in great numbers after the introduction of new, more effective "balloon guns." Balloonists had the most to fear from phosphorus-filled rounds fired from the Vickers .303-caliber machine gun. Ordinary machine guns from a strafing airplane only stitched holes across the envelope, and the slow leaks thus created were not worth the risk of facing strong antiaircraft defenses placed around each observation balloon. But a streak of fire from phosphorus-filled ammunition set the entire craft aflame. Using such a weapon in September 1918, Canadian ace Frank Luke shot down three German balloons in one day. The Belgian pilot Willy Coppens shot down thirty-four balloons in six months. The era of war balloons was over, after 120 years of sturdy service.

This created an apparent vacuum in battlefield observation that would land at the feet of Lieutenant Theodore von Kármán of the Austro-Hungarian Army. Von Kármán had been an aeronautical scientist and engineer before the war. Regarded as a mathematical prodigy at age six because he had been able to solve his older brother's multiplication problems in his head, he had been inspired to pursue the scientific aspects of flight in 1908 after watching Henri Farman win the Deutsch-Archdeacon Prize.[6]

When the army called von Kármán up to serve in the Great War, he had been the young and promising director of an aeronautical laboratory in Aachen, Germany. So the aeronautically gifted von Kármán was pleased when orders arrived in August 1915 to report to the Austro-Hungarian Aviation Corps in Vienna, where he would be research director. The Aviation Corps had a budget that was barely detectable and owned only a few castoff airplanes with which to experiment. But the corps' ambitions rose above all that, which was obvious when in 1917 von Kármán's boss, Lieutenant Stefan Petróczy, ordered him to build a helicopter for use in the war effort. Von Kármán was to start with a model that would remain tethered to the ground by long ropes.

A formidable task, but not an impossible job for a man who would become one of the outstanding aerodynamic thinkers of the twentieth century as well as the head of the Jet Propulsion Laboratory at Caltech. Von Kármán had been the first to work out the math of the alternating rows of swirls that spill off blunt objects moving in a stream of air.* Petróczy asked von Kármán for a *Fessel-schraubenflieger*, which was German for captive helicopter, but the project became known as the PKZ, the initials standing for von Kármán and two of his co-workers, Stefan Petróczy and Wilhelm Zurovec. And when the helicopter began to take shape, that shape was odd.

Today's helicopters always have their rotors at the highest point and all weight hangs from the rotor hub. On the PKZ, the two coaxial rotors were in the middle. The observer's tub sat atop the rotors, and landing gear and the engine took the low position. The observer had a machine gun in his tub for defending against strafing aircraft but there was no room for anything else. The bare-bones PKZ was never intended to zoom around the battlefield. Its only job was to rise high enough so that an observer with binoculars could check out the enemy lines and call back down over a cable with the information. If artillery batteries needed a view from across the hill, the whole support operation of two dozen men and two vehicles had to pack up and move. Three steel cables, each spooled onto a winch, anchored it to the ground. Inflated bags served to soften the landings.

There were many experiments, some of which flopped immediately, like the custom-built electric motors intended to drive the rotors. These caught fire after fifteen minutes of operation. The von Kármán team switched to three 120-horsepower Le Rhone gasoline-fueled rotary engines that they had pulled out of downed enemy aircraft. According to the inventors, the most successful model, the PKZ 2, made at least fifteen unmanned flights in Austria. They said this was enough time to notice that it was at a higher risk of crashing if the engines weren't turning out enough power to keep the cables fully taut, and if the winch operators weren't paying attention to the problem. In that case, allegedly, the PKZ began swooping from side to side like a trapped animal until it tilted too far sideways and came down.**

Petróczy obtained a Hungarian patent for it, but his postwar

---

* The math of vortex formation is uncommonly complicated, but the name von Kármán gave the phenomenon is simple: He called the effect the vortex street. A related phenomenon, vortex formation near the tips of helicopter rotor blades, is the bane of helicopter designers and pilots.

** According to rotorcraft historian Roger Connor of the Smithsonian Institution, there is no evidence that the PKZ 2 ever flew with an observer aboard: "There is a doctored photo floating about showing someone in the observer's basket while flying," he says, "but they never got it stable enough to attempt that." The concept was resurrected briefly before World War II by Germans seeking a way to raise radio antennas.

plan to sell them for flying tourists was never realized. The Italian army hauled the hulk away and put it in a museum. The unusual layout adopted by the PKZ—people above, rotors below—didn't wink out entirely. Other prototypes with low-mounted rotors, such as the de Lackner Heli-Vector, flew in the 1950s.

Publicity about the PKZ led to brisk speculation by aviation thinkers about what else helicopters might do in wartime. In 1917 German commanders summoned von Kármán to Berlin to report on whether a radically improved PKZ might be able to haul supplies across enemy lines and into Berlin if the city were cut off by Allied forces in the way that Paris was surrounded by Prussians in 1870. More speculative was the possibility that a military helicopter could range out to help people, perhaps in the manner that Robur had reached down from the *Albatross* to rescue shipwreck survivors. The great majority of wounded men traveled by truck or automobile to hospitals in the rear, along rutted and shell-blasted roads. Ernest Hemingway described the torment of such ambulance travel in his novel *A Farewell to Arms*.[7]

The U.S. Army caught the helicopter fever also and activated a development program soon after the war. Whom would bring it forth?

On June 15, 1935, Princeton University's International Relationship Club hosted a lecture by the Russian-born mathematician George de Bothezat. De Bothezat, stocky, bald, and with a waxed mustache and goatee, favoring three-piece suits whenever speaking in public, had once declared himself the world's greatest mathematician. Though the claim did not take him far—Einstein, for one, was alive at the time—de Bothezat was the author of many books on a wide range of subjects from economics to engineering.

His talk that day was titled "The Isochronism of Time." It was a frontal assault on Einstein's theory of relativity. De Bothezat was writing a book featuring this point of view, to be titled *Back to Newton*. Many mathematicians were in the audience, including the head of Princeton's math department. According to a bemused reporter for the *New York Times*, a representative portion of de Bothezat's lecture ran as follows: "Once mutual isochronism of time can be proved between two synchronous bodies the self-isochronism of time can be proved in relation to either one through the whole of existence."

When de Bothezat finished and asked for questions in a challenging manner, no hands went up.

Finally a frizzy-haired man in the second row stood. The man was Professor Albert Einstein. Einstein started asking questions in a quiet and thoughtful tone. De Bothezat answered formally, in a guarded fashion. Einstein advanced to the blackboard. Einstein took up a piece of chalk and started writing. Few could follow the details of the showdown, but they could see that something was happening, because de Bothezat was waving his arms and his voice was getting squeaky and shaky. Einstein continued asking his questions and dashing off equations for almost half an hour, all the while wanting to know exactly how de Bothezat had come to his conclusions. De Bothezat finally conceded a point and Einstein sat down next to his wife.

While history does not confirm de Bothezat's self-assessment as "the world's greatest scientist and outstanding mathematician," he did play a meaningful role in the American postwar helicopter program.

The prime mover behind the army's interest in rotary-wing flight was Army Major Thurman H. Bane, commandant of the aviation-research branch at McCook Field in Dayton, Ohio. Bane was looking for ways around the fact that cheap, surplus warplanes were discouraging expensive research into new airplanes; he once grumbled that the best thing that could happen to a fleet of broken-down Handley-Page British bombers was to have them catch fire. On the other hand, helicopter research had no such tattered cloud of war surplus hanging over it. Bane had heard of the Hungarians' PKZ 2 during the war, and had read de Bothezat's 1916 scientific papers on rotor-blade theory. Bane invited him to send a proposal to build a free-flying helicopter for the army. The United States, fretted Bane, was the only major country without some kind of military helicopter in the works, and de Bothezat had a design ready to go. Bane saw no point in broadcasting for competitive bids because, he felt certain, no one in the world knew more about helicopter rotors than de Bothezat. De Bothezat had been a student prodigy of Russian helicopter theorist Nikolai Zhukovsky, who had done leading work in the field, so de Bothezat came with good credentials.

The army accepted de Bothezat's letter as an official proposal and told him to begin construction. Though de Bothezat received most of the attention during the project, his partner, Ivan Jerome, supervised the work.

De Bothezat insisted that the helicopter, code-named Engineering Division H-1, be shielded from snoops. Accordingly, when it had to be moved outside for assembly Bane arranged for a tall screen of black cloth on a wood frame. Pilots of McCook drove de Bothezat into a rage by flying low over the enclosure around the aircraft as if to report back to Moscow, Berlin, or Tokyo. The work went on for eighteen months, seven days a week. Even so, de Bothezat found time to give lectures in the Dayton area about how humanity was going to fly off to other planets aboard rocket-powered spaceships.

As did many early helicopters, the H-1 looked like a monstrosity when finished. Viewed from the air by the annoying pilots of McCook, it was a vast four-rotor helicopter. The frame alone was sixty feet across. It had four arms extending from the core section in the shape of an X. Each arm was built of metal tubing and had a large fan-bladed rotor twenty-five feet in diameter positioned at the end. There were four more rotors for control.

The 170-horsepower H-1 flew beginning on December 18, 1922, and the evidence is clear that it hovered without any help from people on the ground. Thurman Bane, retired from the army and now the test pilot, was able to keep the 3,600-pound machine under control. It was a remarkable piloting achievement, because one pilot said later that the ideal pilot would have as many appendages as an octopus. To hover required handling a stick and rudder pedals like those on an airplane, plus a wheel like that of an automobile, plus another wheel inside that one. After two minutes of hovering at an altitude of ten feet, Bane found that he was 300 feet downwind and coming up on a fence. He landed before tangling with it. Said a favorable witness, "Such performance meant that the smallest city lot, street block, or the top of a building could be used for an aerial station."

A hundred more such flights followed, including some demonstrating low-speed flight. Within six months the H-1 lifted four men, who hung on the crosspieces as if from a jungle gym. But the men were in no danger of a great fall if they let go, as the machine, despite

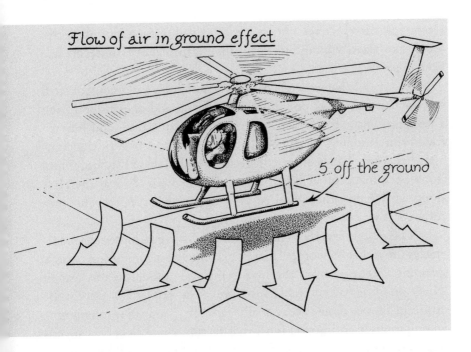

Flow of air in ground effect

5' off the ground

further tinkering and a bigger engine, could rise no more than forty feet off the grass of McCook Field. The H-1 was unable to rise out of ground effect, which is the helpful cushion of air that assists any wing-supported object when it is close to the ground. This cushion goes away at heights greater than about half the diameter of the rotor.

De Bothezat told a reporter in July 1923 that while he admitted the H-1 was not something the army could put to work right away, it should rank in the annals of helicopter history in the same way that the Wright Flyer ranked among fixed-wing airplanes. But the H-1 found no such place in history books. It was too complex to be reliable and was trapped within the original class of early helicopters, which were more floaters than flyers. The army's report grimly recommended not developing any more of them "except in the case of such military urgency that the life of the pilot and observer is of little consequence."

The army had put $200,000 toward the helicopter, which was a great deal of money at the time, particularly in light of the fact that the army's annual appropriations had been shrinking year by year. The entire budget of the base that hosted de Bothezat's work, McCook Field, was only $5 million in the year the work began.

Despite the army's expenditure, the H-1 was not a financial success for de Bothezat, who did not earn the performance bonuses he had been expecting.

Today, all that remains of de Bothezat's H-1 is a control stick and the metal frame of one rotor, which hangs in lowly fashion from a ceiling of Paul E. Garber Preservation, Restoration, and Storage Facility in Suitland, Maryland. Just before World War II, the complete helicopter had been in storage at Wright Field in Dayton, Ohio, along with many other historic aircraft. But with the war the H-1, along with many others, melted away in a scrap drive. To helicopter curators, it was a crime against posterity.

Taken together, the artifacts that do survive in museums and libraries show that helicopters had an extremely long gestation period, spanning 150 years from the first flying scale model to the year in which one was fully suitable for travel. The early ideas were optimistic in the extreme, occasionally farseeing, and often tragicomic when glimpsed in historic films that feature some rickety scaffold-like contraption flying to pieces rather than flying at all. Despite the infrequency of success, many inventors could not shake the fear that idea-thieves lurked nearby and were going to beat them to the patent office. Said one article of 1910, talking of the throng of obsessed inventors who appeared in Hammondsport, New York, during that town's one-year aeronautical boom: "The cranks with the freak machines are all that way. They are all obsessed with the hallucination that the rest of humanity is trying to steal their ideas."

The world's oldest surviving helicopter belongs to the Smithsonian and is on loan to a small aviation museum at College Park, Maryland. To the eye the locale is just another pocket-sized private airport, thoroughly bounded by buildings, trees, and railroad tracks. But to historians the College Park Airport is special: It is the oldest continuously operating airport in the nation. Wilbur Wright trained army flyers here in 1909.

Inside the small exhibit hall is an aircraft with triplane wings, its original lacquer finish browned with age to the color of a mummy's wrap. Wooden rotors top the wings. This is the Berliner No. 5. The chief builder was Henry Berliner, Emile's son. Henry had taken over the helicopter work from his father in 1914. As pilot, Henry began demonstrating distance flights and hovering in front of navy witnesses in 1922.

A television monitor by a guardrail plays portions of a film made at College Park when the No. 5 was new. Somewhat alarmingly, the tips of two propeller-like rotors spin just inches from the pilot's head as the machine hops a few feet into the air. It floats a few seconds, drifting slightly, touches down as if on a low-gravity planet, and rises again. Parts of the machine are even older, since they were recycled from the Berliners' previous helicopter experiments.

Even as real successes materialized in the twenties—mostly in Europe—odd and sometimes interesting ideas clamored for reporters' attention. Edward Fitzgerald's prototype is lost to history because it blew up in 1928 during a test on Long Island; that idea would have put propellers inside ten-foot-high aluminum cylinders and used those for lift. Wilbur Kimball's "helio-copter" of 1908 would have used twenty-four small wooden propellers driven by a fifty-horsepower powerplant. A German named Hanschk claimed in 1922 to have built a helicopter that could cross the Atlantic in one day. Another class was the "helicogyre," in which the helicopter

Curtiss-Bleecker helicopter

rotor blades were made to turn by equipping each rotor blade with its own little propellers. Louis Brennan, of Gyro-Car fame, built such a helicopter and flew it inside a building, then smashed it in 1925. Having propellers pull the rotor blades around the hub avoided the need for a tail rotor, since it avoided a torque reaction, but this came at the cost of much complexity, because power had to come either via a spinning shaft through the hub or by mounting gasoline engines on each rotor blade. The Curtiss-Bleecker helicopter of 1930 was a helicogyre. It was unstable and vibrated dangerously and Curtiss-Wright backed out of the project.

Time and again, reporters passed along the inventors' claims intact and with a breathless tone. "The helicopter has arrived!" announced *Illustrated World* in 1921. "The wingless flying machine is an accomplished fact!" Not until the end of the article did the reader learn that the "thoroughly practical, man-carrying helicopter" featured in the article, the Crocker-Hewitt, had not yet flown.

While most of the prototypes would never satisfy the four requirements of a successful helicopter,* any museum would be proud to hold machines or even remnants from two innovators of the twenties, Raoul Pescara and Étienne Oehmichen. Both created workable helicopters using different designs, and at about the same time.

In 1920 French engineer Oehmichen published a book modestly titled *Our Masters the Birds*. The book was based on three years of intense study of how bird and insect wings work, including high-speed photography. Oehmichen undertook his study while at the French Ministry of Inventions late in World War I, and kept at it after taking a job as an automotive engineer.

Even before his book came out, the wing research prompted him to sketch a helicopter. He convinced his employer, Robert Peugeot of the automobile firm, to pay all expenses. It was of the same basic two-rotor layout Cornu had used, but with rotor blades designed to exploit his knowledge of actual wing performance. Most strikingly, his first helicopter had a sausage-shaped balloon fastened to the top. This gasbag detracted from some of the historical acco-

---

* They must lift their own weight plus payload, hover with good control, move through the air fast enough for useful travel, and be able to land safely with a dead engine. That is, the main rotor must continue to spin without power and produce enough lift on the way down to cushion the emergency landing. This is called autorotation.

lades otherwise due from the aircraft's early free flight as a semi-helicopter, near Valentigney, on January 15, 1921. The blimp envelope was about the dimensions of a modern truck trailer and thus capable of lifting only Oehmichen's body weight. The powered helicopter lifted its own weight.

The blimp was there for one reason: to help Oehmichen keep his prototype under control. Many early helicopters had a nearly irrepressible urge to zoom off in an unintended direction. This "skating" effect still happens when novice pilots like me begin flying modern helicopters, until we learn how to develop a soft but timely touch on the cyclic stick. Oehmichen's balloon served as a brake to these gyrations, but at a cost: it added much aerodynamic drag whenever he tried to fly forward. In 1923 he produced a true helicopter by dropping the balloon and adding eight more rotors to keep control. The frame was arranged in a Maltese Cross, much like de Bothezat's machine. On April 14, 1924, he used it to set the first official distance record for any helicopter flight, by traveling 1,100 feet without crashing.

Four days after Oehmichen set his record, an engineer named Raoul Pateras Pescara de Brindisi bettered the achievement, but with a radically different design.

Pescara grew up in Argentina and he gave up a legal career to pursue aircraft development. He began by designing a seaplane that would launch torpedoes. His seaplane took flight in 1912. Soon after that success, Pescara began pondering helicopters built around a different approach than Cornu and Bréguet had used. Might each rotor blade be put to work doing something beyond just hoisting the machine into the air? Could the blades control the helicopter's horizontal direction so that no other devices were necessary? It was a brilliant speculation. Today we call this a cyclic control, meaning that the pilot can vary the pitch of a rotor blade continuously as the blade goes around, or cycles. Because of the grip that the blades have on the air, the pilot can shift his aircraft almost instantly.

Other experimenters had not yet explored this possibility. Their plans and models treated the lifting rotors like simple propellers good only for lift. They relied on means other than adjustable rotor-blade pitch to give the pilot control over the machine in flight, such as vanes to direct the airflow, or smaller rotors acting like thrusters to shove it one way or the other, or providing a throttle control for each set of rotors. All were proving highly problematic.

In 1916 Pescara tried to interest the Argentine government in paying him to build a helicopter along the new line. The Argentines turned him down, but after he met with Alberto Santos-Dumont, the French military offered support.

"I am going to build a machine which will fly faster than the swiftest airplane, making a speed of 180 miles or more, with which I hope to win the next Michelin race," Pescara predicted in 1921. It was premature: Pescara crashed five machines. Even so he got the cyclic to work well enough that he was able to fly his helicopter inside a dirigible hangar in 1922, which was sixteen years before the famous flight of a German helicopter inside the Deutschlandhalle in Berlin.

Pescara's machine was more horizontally compact than the de Bothezat or Oehmichen helicopters, having the approximate appearance of a roadster topped with a sturdy totem pole, and with that totem pole holding up biplane wings that looked to have been borrowed from World War I aircraft. In one test the rotor hub jammed and the rotors wrapped themselves around the mast, giving the appearance of a badly stowed umbrella. Though the machine looked

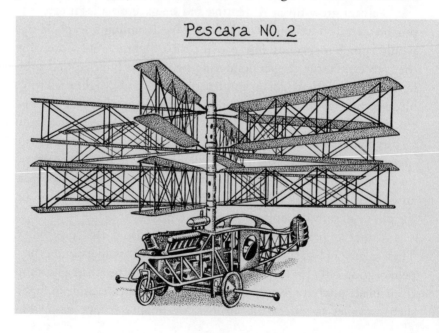

Pescara NO. 2

vaguely comical, many other earlier designs also had a strange appearance. Pescara's machine was the first to demonstrate the potential of a real helicopter.

While a few innovators* looked to be on the way to success, none of their helicopters were ready for action. They were slow in flight and confined to low altitudes. They vibrated intensely and would have had trouble coping with more powerful engines when those became available. In the case of Oehmichen's and de Bothezat's machines, their wide span would have needed too much space to operate under battlefield conditions. Pescara's high-masted design would have been prone to rolling over on the airfield, and it generated much air friction in flight.

These were significant hurdles, but an even bigger one was outside their control. In 1923 something new began pulling attention from true helicopters. Remarkably for a field so dominated by incremental discoveries, the excitement originated with one man, a Spanish patrician-inventor named Juan de la Cierva. Money for helicopter research dried up as investors awaited results. Pescara dropped his helicopter venture and turned to making a race car, the Nacional Pescara.

* To author and test pilot Jean Boulet, three men deservedly share credit for the first controllable helicopters: Pescara, Oehmichen, and de Bothezat. All their machines hovered without assistance from the ground, flew horizontally, and landed vertically.

Chapter Five

Automobile of the Air

Beginning in 1869 and for many decades afterward, Americans were amused by a comic poem written by John Townsend Trowbridge. Trowbridge, born in 1827, was a prolific writer of adventure stories and juvenile novels. The poem was "Darius Green and His Flying-Machine." In the poem the young Darius builds a set of wings, adds a flying helmet and eagle-like tail, and leaps out of a barn loft. He tumbles into the barnyard instead, "as a demon is hurled by an angel's spear / Heels over head, to his proper sphere." Green emerges from the wreckage and concludes that flying is fun but the landing is not. Trowbridge lumped his hero with all those who hoped to flap like a bird; aviation historians have labeled the whole class of such devices under the category ornithopters, meaning "bird wing." The conclusion held by physicists, both before and after the Wrights' demonstrations of powered, fixed-wing flight, was that nobody was going to leave the ground by strapping wings on their arms and flapping.

True enough, but failed ornithoptrians might have found some peace in the fact that the rotor blades on a helicopter must flap, with few exceptions. The effect is dramatic and occurs every time the rotor goes around. The Spanish inventor Juan de la Cierva discovered the importance of rotor flapping while trying to invent an aircraft that couldn't stall its wings and tumble from the sky. His solution, which he trademarked as the autogiro, seemed to overcome some critical shortcomings of helicopters. But the autogiro faced problems of its own, even as it found immediate customers in advertising, personal transport, and news-gathering. In time, the groundwork laid by Cierva and others with the autogiro would contribute much to a new wave of helicopters appearing in the late 1930s.

At its simplest, an autogiro is an airplane fuselage with an unpowered rotor on top, angled to catch the wind when moving forward. The early models came with stubby airplane-like wings to give the pilot control in flight, but in time these were removed as unnecessary. The only wings really necessary for flight were the ones spinning on top. The autogiro's rotor blades are wings shaped in such a way that the air flowing through them produces lift. The job of moving the entire aircraft forward is left to an engine-driven propeller at the front, mounted airplane-fashion.

A high point for the autogiro came rather early in its life, on the New York waterfront in late 1931. Though rudimentary helicopters still had several years to go in proving themselves useful, on December 23 a fully serviceable autogiro was waiting at the wharf on Spring Street when the dapper, smiling Spanish inventor Juan de la Cierva walked off the *Aquitania,* sister ship of the lost *Lusitania.* Reporters were on hand to watch him climb aboard the windmill-like airplane and begin his business travels. It was the first passenger-carrying flight to lift off from Manhattan Island. Cierva flew across the East River to Floyd Bennett Field in the autogiro, then to a country estate in Pennsylvania.

It was a publicity stunt but deservedly crowned twelve years of hard work by Cierva, the first person to produce a useful rotary-winged aircraft. Perhaps the autogiro was to be that long-awaited aircraft that would take the place of the family automobile, as predicted by English businessman Harry Brittain. For a time Cierva's autogiro put helicopters in the shade, at least in the public mind. But in time the autogiro would be itself eclipsed when helicopters finally surged forward, just before World War II.

Juan de la Cierva Codorniu was born in Murcia, Spain, in 1895. His father, Don Juan de la Cierva Penafiel, was a lawyer, poet, and later mayor of Madrid and a cabinet-level minister for the Royalist faction of Spain's contentious politics. While Don Juan was supportive of his son's interest in flying as a hobby, he had little time to spare.* Cierva's grandfather was more significant in nurturing the boy's early interest in flying, which he did by taking the teenager on trips to watch the first airplane takeoffs from Spanish soil. Cierva began building model planes at age fourteen.

Three years later Cierva came across the remains of a biplane in Madrid that the owner of a motorcycle shop had flown and then wrecked. Cierva proposed to build a plane out of the pieces and other salvage. The biplane he completed was one of his own design. He called it the *Red Crab*. Many of his countrymen considered it Spain's first aircraft, and the early success made him confident enough to design and build a three-engine bomber in 1919 for a competition sponsored by the Spanish Air Force. It flew but crashed after the pilot made a turn at too low an airspeed while attempting to show off his flying prowess.[1]

"It smashed because the pilot could not stop when he found himself in trouble," Cierva wrote afterward. "It smashed so completely and conclusively that it shook my faith in the conventional plane and started me in search of something better."

The problem was familiar to early pilots: A low-speed banking turn led to a stall in midair, followed by a plunge to the ground. One sign of a skilled pilot was the ability to make a round trip without crashing. Some instructors trained their novices to avoid the risk by flying straight after takeoff, and then landing and taxiing around to face the opposite direction before flying back to the starting point. The response of most aircraft designers had been to work on shaping better wings, and to achieve a higher speed. Either would give the pilot a safety margin during turns. These were stolid, straightforward responses to the problem of stalls leading to crashes.

But Cierva wondered why airplanes had to fall out of the air. He decided that if an airplane's wings rotated, driven to spin simply

---

* As the morality-minded civil governor of Madrid, Don Juan arrested prominent citizens for gambling and closed all restaurants at midnight. As minister of the Interior and War Departments, he carried out many government reforms and on occasion threatened to shoot his opponents.

by the flow of air, no pilot, however careless (and he had one in mind), would be able to stall the aircraft. Even if his engine quit, he would settle to the ground as if lowered by parachute. Cierva had played for hours with a Launoy-derived helicopter toy as a child, so the idea of freely spinning rotors came easily to mind.

Cierva tested his ideas by building little airplanes topped with rotors and dropping them off a balcony. After graduating from engineering school, he went to work at his father's business. But still he pursued his idea. By early 1922, he had assembled three full-sized prototypes, C.1 through C.3. Each took a slightly different approach in the rotor layout, but all shared one problem: they rolled to the right side in tests, despite the best efforts of the pilot to counteract this by tugging on the controls. Cierva went back to his shelf of scale models. One scale model in particular appeared promising in flight, though the full-scale one identical to it, the C.2, had rolled over like the rest. Pulled along by a rubber-band-driven propeller, the model took to the air, flew forward nicely, and settled softly to the ground. Both the model and the C.2 had a single, five-bladed rotor atop the fuselage. What could possibly be different about the scale model?

While watching a performance of *Don Quixote* at the Royal Theater in Madrid in January 1922, which featured a windmill onstage, Cierva noticed that the blades flapped slightly with each rotation. On the full-sized C.2, the rotor blades couldn't flap, since they were braced by wires running from the hub, in the way that radio antennas are braced from the ground by guy wires. Cierva went home to examine his strangely successful scale model once again. As with *Don Quixote*'s windmill, the blades on the scale model were made of flexible slivers of palm-tree wood, called rattan. The flexible blade of the model made all the difference. Though Cierva would never try to build a true helicopter himself, in time his revelation proved to be a critical discovery that made helicopters possible. He had experienced a lightbulb moment, the single most important discovery in giving rotorcraft freedom of the air.*

Without blade flapping, autogiros as well as helicopters would

---

* Several of Cierva's predecessors had foreseen that blades would need to flap. Charles Renard had mentioned blade flapping; and Louis Bréguet had patented it, as had two Hungarians living in Germany. And Elisio Del Valle's helicopter-like clown toy of 1890 had hinged rotor blades.

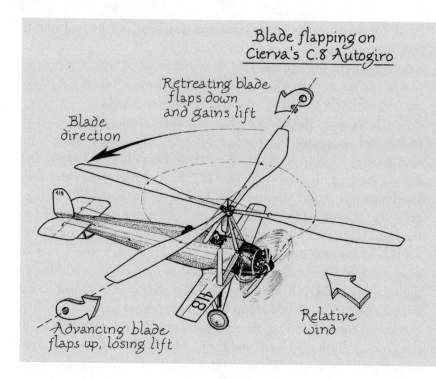

Blade flapping on Cierva's C.8 Autogiro

Retreating blade flaps down and gains lift

Blade direction

Advancing blade flaps up, losing lift

Relative wind

be limited to a slow forward hover. To explain in the context of autogiros: Cierva's early autogiros had rotors that spun clockwise as viewed from above.[2] This meant that the blades on the left side were moving into the airflow whenever the aircraft moved forward; on the right side, they were retreating from it. The result of this "dissymmetry of lift" problem was that the autogiro went greatly out of balance in forward flight, the lift being much greater on the left than the right. This imbalance rolled the machine over whenever the aircraft reached cruising speed. But if allowed to hinge, the advancing blade could flap upward, giving up some of its lift. The retreating blade flapped downward and produced extra lift. In a veritable miracle of aerodynamics, everything balanced out because the blades hunted their own ideal position while revolving.

Cierva's flapping-blade solution remains the key to all single-main-rotor helicopters today, allowing them to fly as fast as 250 miles per hour.* One result of blade flapping is that most of the lift that

---

* Practical top speeds are closer to 200 miles per hour. At high speeds helicopters pitch up, which slows them.

holds up a cruising helicopter comes from the very front and the very back of the rotor disk.

Cierva's test pilot, the capable and remarkably named Gomez Spencer, tried out the new flexible-blade solution for the first time on a fourth aircraft, the C.4. It flew immediately on January 17, 1923. Cierva started building more. His second major advance came with the C.6, an autogiro with improved takeoff performance that he flew often for demonstration purposes. After a rotor blade cracked with fatigue and broke off in flight, he decided that the blades needed a second hinge that would allow them to move forward and backward as well as to flap up and down.

Other improvements would come, but even at this stage Cierva's autogiros could, and did, land on a tennis court. Fame spread when he flew the English Channel in 1928; from that point on, Cierva called England his headquarters. Cierva cut his ties with Spain.[3]

Although he crashed several autogiros during his tours of Europe, Cierva always walked away from the wreckage, and the mishaps seemed to support rather than weaken his argument that the aircraft for the common man had arrived. If not perfectly safe, it seemed nonfatal. He oversaw improvements that moved the machine closer to full helicopter status. One was a clutch arrangement that allowed the engine to spin up the rotor while the aircraft was still on the ground. This shortened the takeoff roll greatly. Another improvement was to give the pilot enough control over the rotor blades that he didn't need the stubby wings and their ailerons anymore.

While the initial inspiration for the workable autogiro at *Don Quixote* may make Cierva seem an amateur with luck and good timing, experts knew better, given his groundbreaking treatise on the aerodynamics of rotary wings.[4] His insights into how a rotor should behave in forward flight had a permanent effect on all rotorcraft. For one, he reduced the breadth of rotor blades to make them more efficient. Oddly, the narrower blades performed better than ones built broadly.* Cierva had uncovered the counterintuitive fact that narrower blades wasted less energy on air friction. Further, Cierva was the first to treat a rotor blade as an airfoil that had its own peculiar needs, distinct from an airplane propeller.

---

* Engineers call this a low-solidity rotor.

Autogiros found only a small clientele, though publicity made the group seem much larger. According to Senator Hiram Bingham, as of 1931 the autogiro was "the hope of the future for commuters of the future." Bingham, best known to the public as leader of the Yale expedition that discovered the lost city of Machu Picchu in 1911, used an autogiro himself while campaigning in 1932. Autogiros helped the New York Police Department with traffic management on Long Island. Admiral Byrd used one during his 1933 Antarctic expedition. Autogiros helped explorers chart the ruins of Chichén Itzá in Yucatan, brought Santa Claus to Van Cortlandt Park in Manhattan, and patrolled forests for Dutch elm disease. One autogiro pilot received a prearranged "parking ticket" from the chief of police in Miami after landing in a park. The pilot smiled for reporters on hand and gave the chief a ride afterward. A radio program in New York for teenagers featured the autogiro. Autogiros brought late-arriving mail to outgoing steamers by dropping it on a rope.

The military experimented with the contraption. In June 1932 the marines loaded one of their jump-takeoff autogiros on the USS *Vega* and brought it to the undeclared war that U.S. Marine Corps expeditionary forces were fighting in Nicaragua. Any hopes of proving the Pitcairn OP-1 autogiro a breakthrough for wartime use were doused in the five-month test, however. With two men onboard the only way to keep the takeoff distance short was to limit any payload to fifty pounds; when 200 pounds of payload were onboard, the pilots considered it unsafe. Absent a strong headwind, it couldn't hover. As a first-generation autogiro it offered the pilots no way to use the rotor for steering the craft. While landing at low speeds little air was moving over the short wings and rudder, and therefore pilots had little control at that critical phase.

Even good landings disturbed some pilots: "The rotors let the OP-1 down with a rocking motion which had the effect of seasickness on some pilots, while making others sleepy," wrote marine historian Lynn Montross in 1954.

Given a new generation of high-payload, short takeoff airplanes, the early autogiro offered few advantages to the military. It could scout from the air, and it could deliver one officer at a time to remote areas. It could land to bring back an injured man, but only if the wounded one sat upright, since there was no room for a stretcher. Back home, it flew the secretary of the navy to join President Herbert Hoover at a fishing camp in Virginia in 1931.

To help open up the Western Hemisphere, Cierva turned to a young aviation executive named Harold F. Pitcairn. His father, John Pitcairn, born in 1841, had founded the Pittsburgh Plate Glass Company. Harold Pitcairn was more interested in aviation than the glass business; by age thirty he was operating airfields and a network of mail-handling routes that was to become Eastern Air Lines.[5]

Pitcairn had begun building powered helicopter models in 1925; one flew out of control once and cut him on the nose, leaving a lifetime scar. After watching a demonstration of the autogiro in 1928, captivated by the possibilities, Pitcairn bought the right to sell Cierva machines in the United States. He sold off his other aviation companies to lay the financial footing for the Autogiro Company of America, which would handle the licensing.

"It is no freak," Pitcairn told reporters about the autogiro upon returning to America in April 1929, "but an efficient product of a sound, conservative aerodynamic engineer." Pitcairn set up a second aircraft company (Pitcairn-Cierva Autogiro Co.) in the hopes of developing a model suited for the American market. After several prototypes, this company came up with the PCA-2 model, which the company flight-tested with good results. Pulled by a 330-horsepower engine, it carried two passengers, seated in front of the pilot. It was the first American-produced autogiro and went on sale in 1931.

Whatever its virtues in the air, the timing on the ground was unfortunate. With a depression under way, at a price of $15,000 the Pitcairn PCA-2 was three times more expensive than a comparable fixed-wing airplane.* One of those to take a flyer on the craft was Johnny Miller, who would be the first private owner of the PCA-2 and also one of the first pilots to master rotary-winged flight, so long ago.

But not so far away. I found him living on the outskirts of Poughkeepsie, New York. He was titled Captain John Miller,

---

* As an example, the Cabin Waco, with an enclosed compartment seating four people and a top speed of almost 150 miles per hour, sold for $4,995.

according to his mailbox. Miller, who at ninety-nine was still an active pilot when I interviewed him, had logged over 30,000 hours as pilot-in-command. That included time as test pilot, barnstormer, aerobatic performer, helicopter pilot, and airline pilot. His memory for aviation details was beyond encyclopedic. He resided in his boyhood home, a former barn remodeled into a blue house south of Poughkeepsie. Airplane models hung from the ceiling of his study.

Miller was thin, tall, and confident, as shown in photos of him alongside historical figures of the early twentieth century; but also white of hair and hard of hearing. Just a short walk across the highway and down the hill is an area once known to locals as Camelot. On May 29, 1910, as a toddler, Miller saw Glenn Curtiss when he stopped for fuel, arriving from Albany in the *Hudson Flyer,* on his way to the history books and Manhattan, where he would win a prize for the first American long-distance airplane flight. Miller's father told John to remember the day. Then age four, Miller resolved to drop his interest in steam engines. He would make flying his career.

By 1923, at age seventeen, Miller was still keen on flying. He saw an article in the *New York Times* about Juan de la Cierva's first successful flights in the autogiro. Miller wrote to Cierva in Spain, not expecting an answer.

"I didn't know his address but it was forwarded," Miller recalled. "I asked for an explanation of the autogiro. He wrote me back, about how it was not a helicopter. So I wrote a second letter and asked for the aerodynamics. He wrote back and explained the aerodynamics, so I guess I was the first person in the U.S. to learn the aerodynamics of rotary wing flight. Here he was, famous, and he had the decency to write to a high-school student."

A year later, Miller became the owner of a worn-out JN-4 airplane, given to him by a barnstormer in return for a summer of mechanic work. The pilot moved on before teaching him what to do with it. Miller read a book titled *Aerobatics* and decided he had enough knowledge to get into the air. So without a single lesson Miller started up the engine and got comfortable on the controls by bouncing and flying low around a hayfield. He learned the rest in the air and began offering rides to local residents in 1923, charging up to five dollars per customer. In 1927 he graduated from engineering school; that was the year he watched Lindbergh take off from

Roosevelt Field for Paris. He began his flying career as a mechanic for the Gates Flying Circus. As a barnstormer he landed at small towns off the traveled air routes, where people would pay a dollar for a one-minute ride.

In 1931, Miller was operating a small airplane-repair business in Poughkeepsie. Business was brisk, much of it fixing airplanes that had suffered hard landings at the hands of bootleggers who taxed them with too much liquor. That year, still intrigued by the possibilities outlined in Cierva's letters of eight years before, he felt confident enough to buy a PCA-2 autogiro from Pitcairn-Cierva Aircraft. Miller immediately set out for the West Coast with plans to exhibit along the way, knowing that Amelia Earhart had the same destination in mind and had allied herself with Beech-Nut Packing Company, a chewing-gum concern that already had an autogiro on order.

Miller completed the country's first transcontinental rotorcraft trip by arriving in San Diego on May 28, 1931, two weeks later. In Los Angeles, he gave a courtesy ride to film director Clarence Brown of MGM. By pointing his autogiro's nose into a stiff wind, he made the autogiro hold position for fifteen minutes over the Los Angeles City Hall.

Amelia Earhart left the East Coast at the end of May by autogiro, bound for California. During the return trip she crashed while attempting a takeoff from an air show in Abilene, Texas. Earhart had another mishap while making a landing in Detroit in September. Her husband put out a press release blaming the craft for structural problems in its landing gear, but federal authorities assigned the cause to pilot error.

Regardless, sales continued. Of the forty-six Pitcairn autogiros sold in 1931, many were purchased by businesses. Silverbrook Coal used one for executive transport. Champion Spark Plugs and Standard Oil promoted products with it. Not yet discouraged by the economic depression, two other aircraft companies signed up to license the autogiro patents.

The expenses were steep for a solo practitioner, and Miller paid the bills by displaying the slow-flying autogiro above the masses. He performed autogiro loops and rolls at an air race in Cleveland, the first time the public had seen such a thing. He dragged advertising banners constructed of rattan and burlap through the sky above East Coast beaches. It was not the job for an amateur, because the banners were heavy and added much drag.

"I did not usually know what they were advertising because I was too busy refueling and getting ready for the next tow to look to see what they were advertising, and it did not make any difference to me," said Miller. "In one case, however, I carefully dropped a banner so that it spread out without getting tangled, then looked and it read, 'Bali Bras, for Yours & Ma's,' and then a phone number to call. It was towed around Manhattan Island. They said that they were flooded with phone calls." The only opposition to his advertising came from the Socialist-leaning authorities of Jones Beach, which banned all advertising within sight of the shore, even company names on delivery trucks. Despite the marginal flying qualities of the autogiro when dragging such a heavy weight, he had only one forced landing during the banner years, when his engine quit in 1934. He cut loose the banner and landed in a graveyard in North Arlington, New Jersey.

In 1934, Miller saw an article about a sportsman called John Randolph Hopkins, one of the heirs to a business empire begun by the inventor of Listerine mouthwash. Hopkins was quoted as saying he thought airplanes were unsafe but would consider buying an autogiro. Miller wrote him a letter offering his PCA-2 autogiro for sale. Hopkins bought Miller's autogiro and also hired him as his personal pilot, a position sometimes called an "air chauffeur" among the landed gentry.

Fairly or unfairly, Hopkins embodied the image that pop culture was giving to the sort of people who favored autogiros as their aerial limousines: wealthy and flighty. The movie *It Happened One Night* featured a would-be bridegroom arriving at his society wedding by autogiro. A cult mystery novel of the era, *The Marceau Case* by Harry Stephen Keeler, set up a murder plot in which detectives guessed that a midget assassin had used an autogiro to make his escape after committing a murder in the middle of a millionaire's croquet lawn. From magazine advertisements and many articles, one might conclude that autogiros were licensed to operate exclusively from country clubs, derbies, foxhunts, Ivy League football games, and other hideaways of the idle rich. According to a Pitcairn magazine ad of the day, "Private

owners not only avoid the crowded highways, but add the fun of flying to their enjoyment of the content on the ground."

Miller housed the autogiro in Great Barrington, Massachusetts,
near Hopkins's estate in Lee. Hopkins was a "youngish and pleasant"
person, Miller recalls, and kept a private zoo at his estate. The estate
had a landing field suitable for autogiro operations and Miller
alighted there when picking up Hopkins and his wife for a trip. The
autogiro had an open cockpit and seated three people. Miller flew his
employers on shopping trips to New York, where they had an apartment on Park Avenue, and as far away as Palm Beach, Florida, where
Hopkins had a winter estate. The couple's 109-foot yacht, Hopkins
told Miller, took too long in making the trip.

In the Miami area the society scene revolved around a set of in-
terlocking resorts known as the Florida Year-Round Clubs. Using
the clubs' autogiro, Year-Rounders could fly comfortably over any
unseemly scenes of Depression-era hardship when shuttling among
the Miami Biltmore, the Roney Plaza, and the golf links.

The standard for such flamboyant behavior as an autogiro
lifestyle had been set for young Hopkins by his forebears. Hopkins's
paternal grandfather, a prominent Atlanta loan broker, had gotten
into a huge and ugly divorce proceeding in 1914 after his wife, a
woman of high society, took a trip to New York. Hopkins's grandfather told a court later that his wife had been completely spoiled by
the city life, refused to come back to Atlanta, spent cash like a mad-
woman, and had put private detectives on his trail. His wife replied
hotly that he was not one to cast stones because he kept their house in
Atlanta blazing with lights throughout each night and entertained
lewd women there, including one named in the papers as Martha
"Bungalow" Harrison. She said the checks that her loan-broker husband sent to pay expenses were so meager that she was a virtual prisoner in her luxury apartment in New York's St. Regis Hotel because
her jewelry was under lien for unpaid bills.

This brawling couple spawned only one child, Russell F.
Hopkins, eventually to be the father of the autogiro-owning J. R.
Hopkins. In 1906 Russell eloped with Listerine heiress Vera
Lawrence Siegrist on the yacht *U-No* while it was anchored in the
Hudson River. This alienated his bride from the Listerine clan griev-
ously. But in time the Listerine money would go far to support

Russell and later Russell's son John Randolph. Russell was an avid collector, focusing on wild animals and every variety of insurance policy. Policies that he bought covered him against burglary, conflagration, tornado, and automobile accident. One insurance policy protected his animals, and another was written to cover an assassination attempt by the Black Hand, a group of extortionist bomb-throwers then notorious in the newspapers. As a young man Russell kept carnivores in the backyard of his Atlanta home till the neighbors complained to the police. Later he built one of America's biggest private zoos at Veruselle, the family's retreat along the Hudson River. He stocked the family house on Manhattan's Fifth Avenue with more wild animals, which lived in a compact zoo on the roof alongside a fenced rooftop playground for his four children.

One of those children, John Randolph Hopkins, was truly a silver-spoon baby. When John Randolph was two months old, Russell gave him a house on Manhattan's Fifth Avenue. By 1928, when John Randolph was sixteen, both parents had died and he inherited enough money from the estate to live the life of the wealthy sportsman. During the two years Miller flew for Hopkins, Hopkins would sign his name on an entire stack of blank checks and hand the checkbook over to Miller. "I was scared to death some crook would get it," Miller said. "I used it to pay the expenses related to the autogiro and my own salary."

Over the years Miller flew for Hopkins, autogiro manufacturers continued hunting for an enduring market, with ingenuity if not success. Citizens took renewed interest in autogiros after seeing the results of a 1933 competition for a safety airplane, organized by Eugene Vidal, head of the Bureau of Air Commerce and the father of author Gore Vidal. As its entry, Pitcairn's autogiro subsidiary produced the AC-35, a snub-nosed autogiro that could cruise the streets at twenty-five miles per hour once its rotor blades were folded back. Propulsion on the ground came from a powered tail wheel. Pilot James Ray delivered it to the Department of Commerce by landing on Pennsylvania Avenue in downtown Washington and driving it there.*

---

* Vidal's original proposal had been that the government would select a winner and mass-produce it at a cost of just $700, or 5 percent of the price of a Pitcaim PCA-2 just four years earlier. Though work on the safety airplane continued until World War II, none of the flying cars moved into mass production.

In 1936 Miller resigned from his job as personal autogiro pilot for John Randolph Hopkins* because it involved more waiting than flying. After working as a pilot for United Air Lines, Miller went back to flying autogiros in 1937 as a test pilot for Kellett Autogiro Company. In off hours he flew demonstration flights for Kellett in the vicinity of Washington, D.C., landing on streets and in parks to give rides to air-minded congressmen.

The Great Rooftop Experiment for autogiros began on July 7, 1939. Miller (along with backup pilot Paul Lukens) flew the mail in the KD-1, five round trips every day, weather permitting, for a full year. The certificated operator of the mail line was Eastern Air Lines, the company formed from the airmail operation that Harold Pitcairn had sold off to go into the autogiro business. Each flight moved up to 350 pounds of mail between the Thirtieth Street Post Office in downtown Philadelphia and the airport at Camden, New Jersey, six miles away by air. "The distance was not the significant factor," Miller said. "The point was to see whether or not we could fly in all weather." It was the shortest airmail route anywhere.

Initial test flights to the 100,000-square-foot post office rooftop had been made by autogiro in 1935, but this was the world's first regular use of a rooftop as a landing field. Miller showed me a home movie of the rooftop comings and goings taken by a friend. The landing in the film is striking: Miller's autogiro approaches at a steep angle, nose high, and touches down lightly on the tail wheel, in a birdlike posture.

The first landing on the post office roof was not so gentle. The problem was turbulence along the roof, which had long penthouse structures on either side of the north–south landing strip. After that close call Miller paced the parapets and flung strips of toilet paper into the air to map the zones of turbulence. He discovered that the conditions were worst when the wind came from the northwest.[6]

On one of the early flights, in July 1939, Miller brought along orchestra conductor Andre Kostelanetz as an unofficial passenger, folded into the luggage compartment. When the experiment ended on schedule, 2,634 mail flights later, with all contract requirements

---

* Hopkins spent his inheritance and was indicted for federal tax evasion in 1956. Authorities found him living as a hermit under an assumed name near Las Cruces, New Mexico, three years later.

met or exceeded, it seemed like a bold step had been made toward Cierva's vision of 1931 when he'd said: "I expect there to be downtown airports in every town and city, some of them built above the rooftops so that autogiros may alight on them and take off again in perfect safety."

While more publicity flights followed, including a landing at the New York World's Fair in August 1940, Miller switched back to fixed-wing airplanes and, later, helicopters. "I flew three thousand hours in autogiros, and did aerobatics with them, and never scratched the paint," said Miller. He noted that the autogiro had a tendency to self-destruction in the hands of amateurs. Many were destroyed in nonfatal crashes because of pilot error, typically when pilots tried to land them like airplanes.

"The autogiro had a steep glide," Miller said. "If you undershot, you should add power, not pull the nose up—otherwise you'd land short. But you couldn't tell the fixed-wing pilots anything."

The civilian market for Cierva-licensed autogiros was over by 1933, victim in part to the Great Depression but also to Cierva's unfulfilled promises of low price and high payload, particularly when matched against a new generation of inexpensive, powerful airplanes that could land on short airfields, such as the Curtiss Tanager and the Fieseler Storch. Meanwhile, helicopters, while still not on the market, were making a slow but steady comeback. Corradino D'Ascanio set distance, height, and flight-endurance records with his coaxial helicopter in 1930. That year the Vatican announced that it would be the first nation to have a fleet of helicopters. They were going to buy three from D'Ascanio, a spokesman said, and they would be immensely valuable for a state so small that it didn't have room for an airfield.[7]

Shortly after D'Ascanio's tests the Soviet Union produced the TsAGI 1-EA, a helicopter little publicized in the West. The Soviet Union failed to set up an officially witnessed flight so no world record was logged, but the machine did lift off the ground in 1930 and over the next three years it managed to reach an altitude of almost 2,000 feet.*

---

* It was a vindication for early Russian experiments with helicopters; a Russian team led by Boris Yur'yev had constructed the first prototype of the modern helicopter layout in 1912, though it never flew.

Another single-main-rotor helicopter flew at this time in the Netherlands, built by A. G. von Baumhauer. His machine relied on a separate small engine to power the tail rotor. It took to the air only in brief jumps, and was barely controllable. Still, it pointed the way to a Sikorsky-style helicopter. Through the rest of the decade helicopters also made successful test flights in France, Germany, Britain, and the United States. Seemingly all over the world, helicopters were suddenly doing things that autogiros couldn't do.*

Juan de la Cierva would not be part of the great shift back to helicopters. He was a highly talented aircraft designer and, given his study of rotor-blade aerodynamics, might have contributed much toward the shift from autogiros to helicopters.[8] Instead he died aboard a KLM flight leaving Croydon Aerodrome in London in December 1936, when the pilot lost sight of the runway's centerline in fog. The DC-2 airliner veered left, smashing through a tennis court, a house roof, and a telephone pole on the west side of the airport. The fuselage ended on its nose, against the wall of 25 Hillcrest Road. It was the same airport from which Cierva had departed by autogiro in September 1928 for his first crossing of the English Channel, and his triumphal tour of Europe.

"A poet of the machine," read his obituary in the *New York Times*.

A final vindication for Cierva's creation came in the courts. The Autogiro Company of America filed a suit in 1949 for patent infringement against the federal government for buying helicopters after the war without paying license fees on the autogiro concepts on which those helicopters relied. (Sikorsky had paid its patent license fees throughout the period for the use of thirty-nine critical ideas, but certain other manufacturers, and the agencies that bought their machines, had not.) The legal fight lasted twenty-six years. It was an epic struggle, as lawsuits go; one witness spent nearly half a year on the witness stand. Confronted by arguments from the government that autogiros had little to do with helicopters, the Pitcairn lawyers

---

* Veteran pilot Lou Leavitt was of the opinion that autogiro backers had encouraged a "well-nigh hysterical wave of publicity" that had the unintended effect of putting it into the hands of wealthy hobbyists, too few of whom took the time to learn the techniques needed. A fallback attempt to solicit military sales only highlighted its shortcomings. In 1943 Leavitt and other industry figures fretted that helicopters might fall into the same trap.

produced an article from Igor Sikorsky saying that autogiro technology had moved helicopter development significantly forward. Pitcairn's successors eventually won $39 million.

Autogiros also should get credit for giving helicopters a boost with the public. In movies, newsreels, and print, autogiros legitimized the notion that whirling rotors could support a full-sized, man-carrying machine. The stage was set: If helicopters could really take people places, now was the time to show it.

Chapter Six

Jules Verne in a Suitcase

The first genuine cross-country trip of any helicopter came on the morning of October 22, 1937, when Flug-Kapitan Hanna Reitsch needed to get herself and luggage from Bremen to Berlin.

The helicopter into which she climbed was the Fw 61, the initials standing for Focke-Wulf. It was the brainchild of Henrich Focke, an airplane designer who had been intrigued by Cierva's autogiros beginning in 1928, and who angered the brand-new Nazi regime in 1933 by refusing to participate in any manufacture of fixed-wing airplanes for war. Focke wanted to work on a helicopter instead, and got one into the air by 1936. It had two rotors on wing-like pylons, slightly tilted inward for stability.

Reitsch had recently finished demonstrating the side-rotor machine at Bremen for a small audience that included Charles Lindbergh, and had another demonstration coming up in Berlin. The fact that she could contemplate such a trip was a major milestone. Until this flight it had been standard practice for all developers to coddle their helicopters, transporting

Focke-Wulf Fw 61

D-EKRA

them by rail between demonstrations so as not to risk a cross-country crash. Helicopters were hothouse flowers, suitable only for trials in the immediate vicinity of the home airfield.

The helicopter closest to the Fw 61 in performance was its recent predecessor, a coaxial helicopter built by Andre Dorand with assistance from Louis-Charles Bréguet, the same Bréguet behind a first-generation French helicopter of 1907. Their *Gyroplane Laboratoire* shook like a "bag of walnuts," but reached a speed of at least fifty miles per hour and as of December 1935 was counted as the world's best helicopter. Still, Dorand had never been able to meet all six of the tough standards needed to bring home a 3.2 million-franc prize from the French Air Ministry. The team dropped the effort in 1938, when the approach of war prevented them from repairing damage to their prototype, incurred during autorotation tests.

When Reitsch planned her trip, the Fw 61 helicopter had been flying free of a training tether for just sixteen months and Reitsch herself had been flying it for only a month. It was enough time for her to know that the Fw 61 was difficult to handle near the ground in gusty conditions. Given the opportunity, it wanted to roll over in a side wind.

As Reitsch was heading east, less than an hour out of Bremen and over farming country, the oil temperature on the Fw 61's 160-horsepower engine started climbing. She needed to land, because when oil overheats it turns to a charred, tarry substance. Reitsch was a social creature and circled a farmhouse to raise some attention rather than heading for the obvious choice, an empty field. The noise brought a dairy farmer outside: the Nazi publicity campaign had not yet trumpeted the news about its helicopter so the farmer was agog. Reitsch signaled that she wanted to land. The farmer waved back and she parked in his garden, inaugurating the world's first residential helistop. Reitsch took a total of three days, four more landings, and some unscheduled repairs to reach Berlin.

Reitsch had once intended to be a missionary doctor, but gave that up in 1934 to be a glider pilot. Later she became a test pilot for the military. She first flew the Fw 61 in September 1937. Reitsch never forgot her first flight, with Focke looking on. "I was hovering motionless in midair. This was intoxicating! I thought of the lark, so light and small of wing, hovering over summer fields," Reitsch wrote in her autobiography, *The Sky My Kingdom*. "Now man had wrested from him his lovely secret."

Over her fourteen-year career as a test pilot in Germany, Reitsch flew dozens of aircraft for the Third Reich. Some of the assignments were dangerous and she did them entirely out of the limelight, but she didn't turn down opportunities to show off her skill. These would come at the 1938 Colonial Exhibition at Berlin's Deutschlandhalle, where Reitsch would become the first rotary-wing superstar. At 11 P.M. on Saturday, February 19, the helicopter rolled from a booth and into the spotlight. Reitsch climbed in, rose confidently toward the roof, cruised down the length of the 250-foot-long space allotted to her, and came back to land.

The flight was part of a revue (complete with palm trees and simulated African village) called Ka Sua Heli.[1] It was intended to thrill Berliners with music and stories about Germany's colonies in the Southern Hemisphere. The spectacle was advertised as representing a trip through the tropics at 300 kilometers per hour. The two weeks of successful flights, shown on newsreels and executed inside a space smaller than the old Madison Square Garden, would prove to be the high point of the Fw 61's fame:* Wrote the *British Journal of the Aeronautical Society* after the exhibition, "The Focke helicopter presents the first successful solution of the problem... [opening] up fresh fields, which up to now had remained closed to orthodox aircraft." The Fw 61 offered good control; it was good at cruising and reaching high altitudes, and it could glide, or autorotate, to a landing when the engine quit. But it was also slow due to aerodynamic drag and at risk of rolling over in a gusty side wind because one rotor gained lift while the other didn't.

The Fw 61 helicopters went on to make more appearances, such as at the September 1938 annual military air show at the Zeppelin Meadows in Nuremberg, but new Nazi helicopters were coming along and both Fw 61s were out of service by early 1942. The Fl 282 Kolibri, a two-person reconnaissance aircraft, was designed to fly from helipads on ships. A key engineer for the Kolibri, Kurt Hohenemser, was Jewish and the Flettner Aircraft Company in Berlin had to connive to get him on the payroll as a "consultant" because of a ban on Jewish employees.

The most impressive in terms of cargo capacity was the Fa 223, named the Drache, for Dragon. Focke meant it as a larger and combat-worthy version of the Fw 61. Over the course of many tests it flew at altitudes above 6,000 feet in the Alps and once climbed to 23,400 feet. It could carry small vehicles, cannons, and nets of cargo weighing up to a ton. Hitler took a personal interest in the work, and occasional plans were laid for putting it into action; one plan would have used the Drache to aid the escape of Benito Mussolini from the Gran Sasso, where he was being held prisoner by fellow Italians. Nevertheless the Fa 223 never moved from prototype demonstrations into combat. The production target of a hundred machines was blocked at every juncture by materiel shortages on the one hand and Allied bombing on the other. Allied bombers destroyed the first ten in production, then destroyed more with raids in southern Germany in July 1944. In the end only five Draches made it out of the factories intact; the few that survived the war were dispersed to the Allies.

After the war, Focke was as hard up as any other German civil-

---

* However, the crowd's reception to the Fw 61 was cooler than Reitsch had hoped. Apparently the customers had expected from the posters that she would be flying her aircraft at 300 kilometers per hour inside the building.

ian, but occasionally he got care packages from a benefactor with a special bonus sealed in the coffee cans: they contained cigarettes, which were a prime commodity for bartering for food among the ruins of Germany. The boxes Focke received were coming from a Russian-American immigrant by the name of Igor Sikorsky.

In the autumn of 1958, with the aid of binoculars, Igor Ivanovich Sikorsky could have been glimpsed high in the air over the Housatonic River Valley of Connecticut. At the time he was riding aboard a helicopter that a novice might have seen as some kind of aerial illusion: seemingly the sides of the cabin had fallen away and left the four passengers seated in what remained of the cabin: windblown, perhaps nervous, but with a spectacular view.

It was actually a test flight. The helicopter was the Sikorsky S-60, a twin-engine, piston-powered helicopter. The cabin was missing because it wasn't supposed to be there: the S-60 was a "flying crane" designed to lift cargo pallets or passenger pods fitting neatly into the spot where, ordinarily, a cabin would be located. It was carrying a wooden platform and was undergoing flight testing to check for vibration. Sitting on the platform were three engineers and Igor Sikorsky, the quietly exuberant engineering manager of the Sikorsky Division of United Aircraft Corporation.

This was a man who loved aircraft with a view. The rush of air over the Housatonic must have brought Sikorsky's thoughts back to his aviation origins, back to an open-air balcony on the front of a history-making machine that meant as much to him sentimentally as any aircraft he ever created. Its success became his success. Aptly called the *Grand,* it was the largest aircraft in the world when it started rolling down the St. Petersburg airfield on May 13, 1913. It was also the world's first four-engined aircraft.

Even though his work did not occur in isolation, Sikorsky did more than any other single helicoptrian to drag the vertical-flight machine from the experimental workbench and put it to full use. Helicopters formed the third of his three careers in aviation.

Sikorsky was born in 1889 in Kiev, ancient capital of Little Mother Russia. As a boy he built model aircraft and electric motors; he assembled a rubber-band-powered helicopter; as a teenager he mixed explosives based on a recipe from an anarchists' manual and detonated them in the garden. Ivan Sergeiovich Sikorsky, Igor's father, was a man of remarkably wide interests who accumulated a vast private library and was a professor of psychology at the University of Kiev. Igor's sister Olga recalled their father as a man of such simplicity and self-discipline that he refused to sit on padded chairs.

As a bright young man from an academically minded family of good social standing, Sikorsky was on track for a respectable technical career in Tsarist Russia. In 1903 he enrolled at the Imperial Russian Naval Academy in St. Petersburg, where his classmates noticed that Cadet Sikorsky lacked any fear of heights when up in a tall ship's rigging. He resigned from the academy after three years and studied engineering at the Polytechnic Institute in Kiev. He relocated to Paris to avoid social unrest in Russia, and then came back to Kiev.

By 1907, when his mother died, Sikorsky was showing signs of wanting something different from his birthright. He toured Germany the following summer with his father, a time when Wilbur Wright was visiting major European cities by air. While staying at a hotel in the Tyrolean hills Sikorsky assembled a helicopter rotor that was four feet in diameter. He tested it in the hotel room by using weights and pulleys and was amazed by calculations that suggested it would lift eighty pounds per horsepower. He investigated further and realized that this performance would hold true for full-sized helicopters only if they had absurdly large rotors.*

Upon returning to Kiev, Sikorsky embarked on his lifelong career. He quit the Polytechnic and took up temporary residence in Paris, where he learned the basics of flying and aircraft construction from pilots Louis Bleriot and Ferdinand Ferber. Ferber offered the young man this career advice: "To invent a flying machine is nothing; to build it is little; to fly it is everything." Sikorsky also examined Cornu's helicopter in Lisieux. Sikorsky returned to Kiev in 1909 with an Anzani aviation engine and the idea to build a helicopter in his father's backyard.

He was going on funds he had borrowed from his older sister Olga, among others.[2] Igor's aerial infatuation would have been no surprise to his mother, Zinaida, had she been alive at the time, be-

---

* As boys the Wright brothers had been similarly misled by the astonishing performance of their small, rubber-band-powered Penaud helicopters.

cause she had been the first to interest Igor in flying. As a young woman she had been enthralled by the flying-machine sketches of Leonardo da Vinci and had fully conveyed that excitement to her son. She watched as young Igor took up the aerial cause, building gliders and poring over *Robur the Conqueror*.

Upon completion, Sikorsky's prototype H-1 helicopter was a coaxial machine, and it had an appearance somewhat similar to the one then being pursued by John Newton Williams in Washington, D.C., Williams nearing the end of his career, Sikorsky at the beginning. There were two rotors topping a platform, one fifteen feet in diameter and the lower one sixteen feet. These turned at 160 revolutions per minute in opposite directions. When it was stopped, Sikorsky could change the pitch of the rotor blades by tightening steel wires with a turnbuckle.

By 1910 Sikorsky acknowledged to himself that neither the H-1 nor the H-2 prototype was going to get off the ground. Even were the power available, the helicopter lacked control and stability and therefore could not go anywhere. Still, he verified that the coaxial rotors were developing enough lift that they could have thrown the would-be aircraft onto its side.

He shelved the plan, awaiting a future that held bigger engines. He turned the remainder of his borrowed capital toward fixed-wing aircraft. To test propeller designs, he attached the Anzani engine and a propeller to a sled that winter and took to the icy streets. In 1912 he won a major prize in Moscow by setting a speed record with his S-6, a single-engine passenger plane. This led to a job offer from the Russo-Baltic Railway Car Company. Sikorsky paid his family back and now his three-year project looked more like inspiration than a juvenile fancy.

Despite the company's ironclad name, suggestive of railway coaches rumbling across the steppes behind a smoky engine, the Russo-Baltic had enough ambition and money to be a force in airplane development. With its support Sikorsky continued his work on light planes. His S-5 was successful in its way but later crashed as he was flying it, due to an engine stoppage caused by a mosquito carcass in the carburetor. Sikorsky climbed out of the wreck and sat on a pile of railroad ties, reflecting on what to do to minimize the risk of engine failures. More engines would help because they were unlikely to quit simultaneously. He decided to build a giant plane with four engines, the world's first. Russo-Baltic told him to go ahead.

Outside experts could well have judged this an unwise investment, considering Russia's late start in aviation. Instead the project would lift Sikorsky far above the ranks of other strivers, who were still working with flimsy biplanes that a few men could have picked up and carried into a hangar. In six months his *Grand* took to the air, astonishing those spectators who had come to watch it crash. Not only did it fly under good control, but Sikorsky had also broken with convention to give it a long, narrow set of wings that proved highly efficient at low speeds. Aerodynamicists now say such wings have "high aspect ratio." A modern equivalent is the distinctively slender wing of a high-performance glider. Prototypes usually are rough and ready when it comes to comfort, but this one had an enclosed cabin that anticipated the airliner. The ever-considerate Sikorsky also installed tables and chairs on his prototype. The pilot and copilot sat side by side, for the first time on any airplane. The nose had a balcony for passengers who wanted to enjoy the view.

Three months after the first flight of the *Grand* a Polish biplane upon coming apart in flight destroyed it at the airfield, but Sikorsky had already moved on to a bigger version, the *Ilia Mourometz*. The name referred to a tenth-century warrior who loomed famous in Russian folklore. It could carry sixteen people and its wings spanned more than a hundred feet. To a writer from the *New York Times,* the *Mourometz* was more of a "flying village" than a simple airplane. In 1914, the *Mourometz* made a 1,600-mile round trip from St. Petersburg to Kiev. At one point on that trip, Sikorsky found himself forced to fly in zero visibility, but then the airplane rose above it all. He went outside to stroll the balcony in exultation as a carpet of cloud rolled by. With the advent of war there was no need for an airliner, so the production-model airplanes saw service as heavy bombers for the Russian Air Force. They completed 400 missions against the German lines, with just one combat loss.

By March 1918 Sikorsky had assembled an impressive portfolio, but it had no value in the midst of the Bolshevik uprising. He was a "White Russian," whose upper-class background tied him to the Tsarist regime. Sikorsky left for England by steamer with a few hundred British pounds, hoping to restart his aviation work. The war was still under way, and in France he received approval to build a twin-engine bomber that could carry one-ton blockbuster bombs. But the war ended before his prototypes were fully complete. Sikorsky left for New York on the steamer *Lorraine*. He arrived March 30, 1919, stepping onto the wharf at Ellis Island with a single suitcase. In that suitcase, along with clothes, photographs of past airplanes, and plans of those to come, was his boyhood copy of *Robur the Conqueror* in the Russian language.

"The United States seemed to me the only place which offered a real opportunity in what was then a rather precarious profession," Sikorsky wrote later in his autobiography, *The Story of the Winged-S*.

But it did not seem so at first. Sikorsky spent eighteen months trying to start a new company to build a bomber he called the Battleplane. But the postwar market for new airplanes had collapsed under the weight of hundreds of surplus ones. He traveled to Dayton, Ohio, to meet with Colonel Thurman Bane, but the most he could get was a small design contract, which soon ended. By the fall of 1920 Sikorsky, once a star of international aviation when it came to heavy bombers and transports, had washed up back in New York. He moved out of his hotel and into a tiny tenement on New York's East Side.

Spurning eggs and meat as too expensive, Sikorsky lived on baked beans and coffee at eighty cents a day. Not yet fluent in English and having no professional skill outside aviation, he pondered how to make a living otherwise. "I experienced the bitter taste of the cake of fame," he recalled later, in an interview with the *New York Times*. "I was known in my chosen field, but there was little for me to do in it."

Postwar New York suffered from some bitterness as well, arising from the aftermath of the Great War. This was particularly true in the Lower East Side where Sikorsky had taken lodgings. Attorney General Mitchell Palmer was pursuing agitators and to that end set off a wave of anti-Bolshevism. Unemployment in Europe was driving shiploads of immigrants toward Ellis Island.

Some Russians were particularly upset about the deployment of American troops to Russia in 1919, which was part of an unsuccessful attempt to restore pro-Tsarist Russians to power. Bolshevik agitators on the street passed out leaflets daily. In October 1921, firebrands like "Comrade Kaplan" and "Comrade Bob" gathered in Rutgers Square to call in English, Russian, and Yiddish for revolution against capitalistic government and for the establishment of a Workers' Soviet Republic in the United States to replace it. They called for the release of Bartolomeo Vanzetti and Nicola Sacco, two Italian anarchists jailed for murder in Massachusetts.

None of this helped the job situation, but the sturdy network of fellow Russian émigrés in New York led Sikorsky to his first break. A friend told him of a possible job teaching math to laborers at the Russian Collegiate Institute. Sikorsky got the job and added evening lectures on astronomy, which he offered to a broad cross section of ex-Russians. He borrowed lantern slides from the Museum of Natural History for his lectures, and lugged the heavy projector to each gathering via subway. He found the students eager, almost desperate, to acquire a liberal education in history, arts, and literature. Among the students were penurious White Russians like himself but also former Bolshevik sympathizers. "Great credit is due these men," Sikorsky said, "who were willing to spend their money to pay for lessons, the subjects of which were of little practical assistance in their every-day life or work."

His astronomy lectures were well received, but Sikorsky was only a hobbyist in the field. Tell us about airplanes, his audience asked, so he began to talk about aviation as it stood on the brink of a new decade. On this subject he spoke from the heart. He told them about giant airplanes to come. He had been visiting airports in the New York area, where he had watched the bustle with interest and an occasional trace of resentment. He had been out of the business since 1918, so to freshen his lectures he spent days in the public library to catch up on the latest developments in engines and aerodynamics.

Inspired by his talks, fellow Russians encouraged him to build another airplane. He pulled out the old plans from his suitcase and began drawing up an improved transport. By March 1923 Sikorsky had collected enough money from fellow immigrants to begin again—including $5,000 from concert pianist Sergey Rachmaninoff, whose name came to be enscribed on the books as vice president. Stock was priced at ten dollars per share of preferred stock. Sikorsky made it a point to spread the risk among his blue-collar investors and he warned them at every stockholder meeting that the plan of building a multi-engined transport on the Long Island chicken farm of

Victor Utgoff was going to be a long shot. Money was so tight that materials came from military surplus, junkyards, or the local Woolworth store. When mechanics needed to attach the landing gear, they dug a big trench under the fuselage, as no money was available for a crane or jacks.

Immigrants from Russia provided the labor at fifteen dollars per week, as well as the capital. Sikorsky had inspired them with his vision of a Russian aerial renaissance, so at the financial low points they were satisfied with nothing more than meals on the job and stock in the Sikorsky Aero Engineering Corporation. Some even lived at the "factory," which was a farmhouse with outbuildings. These yeomen of the air would be critical in seeing Sikorsky through very difficult times building airplanes and then helicopters. One key figure was Boris Labensky, who served Sikorsky as a design engineer.

Labensky's Russian background was the stuff of company legend. According to Sergei Sikorsky, Igor's son, in February 1919 Labensky had been commander of an Imperial Navy destroyer anchored in the port of Odessa. As the Bolshevik columns closed in by land, Labensky offered his sailors a choice: They could leave in the ship's liberty boats, or they could accompany him and the ship into exile. Half the crew accepted his offer. Labensky steamed down the Black Sea coast and sold the ship—lock, stock, and gun barrel—to the pasha of a Turkish town. He divided the cash and took his fair share. Labensky emigrated to the United States and worked for a time as a tooling engineer. When he first connected with Sikorsky in 1922, he was working on a farm in upstate New York where all the sons of toil were ex-Russian navy officers.

By April 1924 the corporate secretary of Sikorsky Aero warned his boss that the investors were restless: the airplane looked ready to go but hadn't flown an inch. There were rumors that it would never fly. Therefore, whether it was safe to fly or otherwise, the chief executive officer had better get in and fly it around. So Sikorsky did, and so it was that on May 4, 1924, the Sikorsky S-29, laden with so many immigrant dreams, suffered an engine failure and crashed into a ravine on the Salisbury Golf Club course near Roosevelt Field. Eight workmen rode as passengers. Sikorsky hadn't had the heart to tell them that passengers are not appropriate on initial test flights.

No one was mortally wounded except, it seemed, the company.

Even before the crash the firm had been so much on the edge it hadn't
been able to buy new tires for its flagship or even to fill the gas tank
all the way. Sikorsky's men patched up what they could by scroung-
ing more deeply than ever. "Their friendly loyalty made this possible," Sikorsky recalled in his autobiography, "despite the absence of
money for salaries." The engines were beyond patching, however,
and only replacements could put the airplane back in the air.

Sikorsky had a plan. While he had a courtly, Old World demeanor, it was not the same as meekness. He summoned his most
loyal supporters to a conference room at his lawyer's office in
Manhattan, then locked the door and refused to let them out until
they produced money totaling $2,500. He assured them that the risk
was small because he knew from the brief flight that the airplane
would fly well with good engines. They believed him, wrote checks
to win their liberty, and the S-29 took flight as promised that year.

The company's cash reversed its flow for the first time. The
ship's first job was to fly two grand pianos to Washington, D.C. By
July 1927 the S-29 was reported to have flown a half-million miles
and was on lease to a tobacco company as a "flying cigar store." It was
robust enough that when a wing smashed through heavy limbs during a forced landing in 1925, the airplane needed no repairs beyond
a fabric patch. Sikorsky kept the cherry-tree limb and it is still on
display at his old office at Sikorsky Aircraft Division of United
Technologies Corp., a space that is preserved as Sikorsky left it on the
day of his death in October 1972. The S-29 ended its career during
the production of the Howard Hughes movie *Hell's Angels,* modified
to look like a German bomber.

There were more trials to come for Sikorsky's airplanes, the
most traumatic being the fiery crash of the S-35 transport in the first
moments of Captain Rene Fonck's 1926 attempt to cross the
Atlantic. "Ever forward" was Sikorsky's message, and the company
secured its first commercial success two years later with the small
S-38 flying boat, suitable for adventurers and island airlines. Improvements followed, leading finally to the majestic, ocean-crossing
S-42 flying boat. Other models were coming from Boeing and
Martin in the United States, as well as builders in Europe. The short
but romantic era of the flying-boat airliner, swooping from one exotic port to the next, lasted until the start of World War II.

After the Bréguet-Dorand coaxial helicopter began setting helicopter records in France in 1935, Sikorsky took notice. He had dabbled in the helicopter field from 1925 onward, but the transport and flying-boat business (sold by investors to United Aircraft in 1929) left him no time to bring anything with rotors to fruition. Now he set up an office next to the attic of his farmhouse, equipped it with bookshelves and a desk from a rummage sale, and began sketching helicopters again, by night and weekend. He began making day trips to the New York Public Library, the Aircraft Manufacturers Association library in New York, and the Yale University Library. He read all the latest results from Europe. Then he went back to his calculations and altered his earlier designs once more. The timing was good because Sikorsky's flying-boat business couldn't last. By 1938 it was clear that no flying boat, however cleverly designed, could get around the fact that its cruising payload was greatly limited by constraints on payload at the time of takeoff, which were driven by the need to break free of the water. Meanwhile, big, fast land-based airplanes were taking over key routes and there were rumors around the factory that United Aircraft might leave the business to Boeing and the Glenn L. Martin Company. Unless Sikorsky jumped from this sinking raft to a new one, the team of designers and craftsmen he had nurtured from the S-29 days was going to be fired or transferred to warplane development, most likely the Corsair fighter. If that happened Sikorsky would lose any chance of building the helicopter of his dreams.

In 1938 Sikorsky went to Germany to present a paper on flying boats. While in Bremen he saw the Fw 61 in flight. He returned to Connecticut convinced that all the essential elements of a fully usable helicopter were coming together; therefore anyone who hoped to be a player must act quickly. Sikorsky began attending gatherings of helicopter inventors at the Franklin Institute in Philadelphia. Each was alert to the keen competition then getting under way. In December 1938 Sikorsky's sketches were complete. He traveled to Hartford, Connecticut, to meet with United Aircraft's executives to sell them on two ideas: First, the time was right to invest in helicopter development. Second, his patented ideas were the way to proceed. "The presentation went surprisingly well, granted that many in the scientific advisory board were very skeptical of helicopters and were utterly skeptical of Igor Sikorsky's proposal for a new configuration," says Sergei Sikorsky.

The second point, that United Aircraft should follow Sikorsky's design, was not at all obvious. All the helicopters that had been setting records recently were side-rotor or coaxial designs, but Sikorsky wanted to build a helicopter with a single main rotor and a tail rotor to steady it.

The problem was that no single-rotor helicopter had flown even half as well as the side-rotor helicopters, even though experiments on the single-rotor design had begun in Russia in 1912. In what may have been the greatest performance of his life, Sikorsky convinced the executives that the single-main-rotor helicopter (unproven as flyable) had fewer drawbacks than side-rotor and coaxial designs (proven as flyable). He pointed out the Fw 61's tendency to roll over and the aerodynamic interference that occurs when two lifting rotors operate close together in a coaxial configuration.

United Aircraft told Sikorsky to go ahead and work out design details, but refused to approve construction of a prototype until news came later that a little money might be available from the Dorsey Bill, a $2 million allotment toward experimental aircraft designs.*

The prospect of the Army Air Corps handing out $2 million—the first support ever offered to American rotary-wing inventors at large—attracted a horde of hopefuls with designs in their drawers and models in their basements. One application came from the river community of Beardstown, Illinois, where two grocers named Francis and Russell Halligan had gone from building rubber-band-powered models to a full-sized rotorcraft that (they claimed) had flown around town in 1932. Army officer Frank Gregory traveled to their family's home to check it out. As an introduction to the "Halligan Plane," one brother played a piano while the other released a scale model to rise up and skitter off the ceiling. After this overture Gregory was admitted to a back lot, where the prototype was se-

---

* This would be at the expense of money originally expected by autogiro enthusiasts, who had gotten the bill moving in the spring of 1938. The timing of the first Dorsey Bill hearing in April was unfortunate for autogiros; as millions of Americans already knew from the newsreels, the Fw 61 helicopter had recently flown well in Berlin, and a navy official at the hearing had bad things to say about autogiro performance. Congress took out the word *autogiro* and expanded the bill's scope to "rotary-wing and other aircraft."

creted. It resembled an airplane with two rotors for lift, and was powered by a thirty-two-horsepower outboard motor converted for the job. Whether or not it lumbered into the air that day, Gregory struck the Halligan Plane from his short list. He wrote that the craft looked to have been "put together with stove bolts and cast-iron piping."*

To the irritation of all other applicants, the army initially declared Platt-LePage Aircraft as the winner of $200,000, which was most of the sum then available for helicopter work. The army expected the company to build a side-rotor helicopter at least as airworthy as the Fw 61 flying in Germany. Platt-LePage's first helicopter was the XR-1, and test flights began in May 1941. But the company test pilot, Lou Leavitt, was so concerned about control problems when making turns in the XR-1 that he refused to accelerate it to cruise speed and fly the one-kilometer course the army required. Two years passed before an exasperated Frank Gregory climbed in and made the first high-speed flight in the Platt machine himself. The XR-1 later crashed and was repaired; Platt built a second one, but it crashed also. Though it was rebuilt, the army lost all hope by 1944 and canceled the project without seeing more than two prototypes.

Early signs of trouble at Platt convinced the army to hedge its bets by putting $50,000 of Dorsey money toward the Sikorsky Division of United Aircraft. Sikorsky's chosen approach, the single main rotor with a small rotor on the tail to keep the aircraft from spinning, seems obvious now, but many problems, from small to large, needed resolution. Work on the prototype began in early 1939, but years passed before any helicopters were ready for action. They weren't ready in 1942, when the Royal Navy needed a tough, ship-based helicopter that could act as a picket against lurking submarines in the mid-Atlantic, where land-based air cover could offer no aid. During 1940 and 1941, German submarines sunk ships three times faster than the Allies could build them.**

* While the Halligan Plane may have lifted off the ground briefly it didn't impress the army. Interest among unschooled amateurs continued strong for years. In 1941 the aeronautical laboratory at Wright Field, Ohio, set up a Jules Verne Department to process ideas for new air weapons. One entry sketch concerned the Avolater, a craft topped by four rotors. The inventor wrote in marginal notes that it was in a "class by itself" and might reach 1,000 miles per hour.

** The British were so desperate for convoy air cover against the U-boat menace that they considered launching Hurricane fighter planes from ships, using them for a few hours until the fuel ran out, and then ditching them.

However desperate the need, at this early state of knowledge, the development of helicopters could not be rushed. Progress came in slow steps, staying close to the ground and working up from there. The first step—actually begun a year before the company even authorized the work—was to put together a test device with engine, transmission, and main rotor. While not intended to take off, the test rig allowed measurements of rotor performance. The next step was to assemble a flyable prototype. The first VS-300 helicopter[3] was ready for engine runups after just six months, on September 14, 1939. The seventy-five-horsepower Lycoming engine roared as witnesses knelt to see whether all the wheels were going to lift off at the same time. Once free of the stability that the ground offered, the helicopter vibrated so energetically that witnesses said it was "one big blur." Boris Labensky stood ready to tackle the helicopter if it showed signs of zooming out of control.

Sikorsky, as chief test pilot, decided that it would be months at best before he could control the angle of the aircraft by altering the pitch of each main rotor blade, as he had originally planned. The problem was in the cyclic control, which was supposed to tip the helicopter to fly in the desired direction with the touch of a lever. Movies of the test flights were slowed for public viewing to make the helicopter look to be more under control than it really was.*

Sikorsky put the cyclic control aside as a good idea but before its time. As a temporary alternative he had the team add two small rotors on the tail by hanging them off outriggers, like little crane booms. These hampered forward flight, but they allowed Sikorsky to steer the helicopter well enough to sort out other difficulties. Better yet, this success convinced the army to cover some of the troubleshooting costs. Sikorsky set a world record for hovering flight in May 1941 in this "heli," which he pronounced "heely."

Persistent stability problems could have led to daily crashes, but a simple expedient made such testing practical. When aircraft mechanic Bob Kretvix joined the Sikorsky helicopter project in September 1940, one of his jobs was to serve as a human anchor to help keep the great man alive. He gripped a rope passed through a

---

* Years later, Sikorsky commented that he didn't like to reflect back on the risks he had taken during those early test flights.

metal ring on the pavement and connected to the base of the helicopter whenever Sikorsky lifted off to try out the latest experiment. At the first sign of trouble Kretvix and the others hauled the craft back to the ground.

Vibration was a persistent and dangerous obstacle to progress. The big rotor blades had to be exquisitely matched in lift, drag, and weight along the entire length, or they would go around on separate paths and cause the helicopter to lunge hundreds of times per minute. Some problems were simply strange. After watching a demonstration film, a company vice president called to ask why he hadn't seen the helicopter fly forward in the film. Sikorsky reassured him that this was a minor problem—but it wasn't. The reason was a violent oscillation that started at speeds above thirty miles per hour, as if the machine insisted on staying in the slow-moving mode of early helicopters such as de Bothezat's H-1. Sikorsky even contemplated rigging the machine to fly backward to see if that would help diagnose the problem.

"It was okay if he was hovering or flying at ten or fifteen miles per hour, but when he tried to accelerate to twenty or thirty miles per hour," says Igor's son, Sergei Sikorsky, "it seemed that you couldn't fly forward."* Fortunately, the fix for this problem—called the lead-lag damper—would be of assistance in controlling a second threat, which is a helicopter's tendency to self-destruction from out-of-control vibration under certain conditions. That threat was ground resonance. The damper was similar to a shock absorber on automobiles, but it had to be mounted a certain way on the rotor hub.

The team encountered ground resonance for the first time in October 1941. The incident is instructive about what the early experimenters were up against; it also shows how quickly things can go wrong in any running helicopter, experimental or not. At this time the prototype was on rubber, air-filled pontoon floats, and Sikorsky's test pilot Michael Gluhareff brought it in for a landing near the hangar. "As the pontoons gently touched in an excellent soft

---

* The vibration turned out to be an annoying by-product of the fact that the rotor blades on a forward-moving helicopter must flap upward as they head into the wind, and then must flap downward as they head out of the wind. Unavoidably, this up-and-down motion changes the diameter of the swept area and therefore the blades' angular momentum. Each time a blade went around the hub it alternately lunged forward and then lagged back. Lead-lag dampers kept the swinging from reaching dangerous levels. Helicopters use them still today.

Sikorsky VS-300

landing," wrote Les Morris, another test pilot, "the most extraordinary things began to happen." The helicopter began bouncing on the floats, side to side, throwing Gluhareff to and fro, in synchrony. Later the experts would call it the "self-excited instability of hinged rotor blades;" pilots called it ground resonance.*

Gluhareff pushed down on the collective in an attempt to pin the aircraft to the ground but accidentally caught his sleeve on the throttle. This made things worse and he pulled up on the collective to free himself from the ground. He rose, the shaking stopped, he shut off the engine, and the helicopter dropped so forcefully that the tail rotor hit the ground and broke off. The solution for ground resonance and for the dangerous vibration at higher speeds was to make sure each rotor blade was equipped with a lead-lag damper, weighing only a pound. This damper discouraged the rotor blades from swinging excessively on their hinges.

---

* Ground resonance can start if a helicopter begins bouncing on its landing gear and the main rotor blades begin swinging from side to side at their hinges on the hub. It gets worse quickly if the bouncing on the gear falls into harmony with the rotor blades' tendency to swing on their hinges. The result is a rapidly oscillating "center of lift" analogous to an off-center washing machine. In seconds the situation can go violently out of control and ends seconds later when the aircraft rolls over or the rotor system flies to pieces. Ground resonance was the most common reason that Army Air Corps autogiros were destroyed in use.

Throughout the years of vexation, Sikorsky was unfailingly polite, never raising his voice, yet even so, exerting much persuasion. One reason for his persuasiveness was that his men were completely loyal. "He was the most courteous, quiet person," recalls Kretvix, "the nicest guy." He had a way of gaming the personnel system so that when United Aircraft was trying to cut back on manufacturing costs in the flying-boat business, the company found it quite impossible to fire any of his valued assistants. "This story of his moving employees around was more than a legend," Sergei Sikorsky says. "His work force was like a big rubber balloon." Eugene Wilson of United Aircraft recalled later, "A shop foreman laid off one day, for instance, bobbed up the next at a drafting board, and still later as a lathe hand."

Sikorsky knew every man in the factory by name, addressing each with a respectful formality in his Russian-accented English, usually with a slight bow and a click of his heels. He favored hats of all descriptions. Wilson was often amused by the sight of the imperturbable Russian at the controls in the open cockpit, wearing his business suit and a hat that could vary from a golfer's cap to a Homburg to a fedora. "He looked like a baffled professor trying to remember his notes, and I marveled that he ever learned to fly the thing," said Wilson. Among fellow pilots, though, Sikorsky was seen as a test pilot of rare skill on both fixed-wing aircraft and helicopters. Says Sergei, who flew with him often, Igor needed only a brief checkout on an unfamiliar model before he could fly it like a regular. Sergei put it down to the skills that Igor had developed in his earliest days of flying, as a means of survival.

"Igor was fearless, the original Fearless Fosdick," says Harry Nachlin, who worked for Sikorsky during the war as a mechanic. He was referring to a comic-strip character created by Al Capp. "He had crashes but walked away from all of 'em."

By fall 1941 the helicopter was flying well enough that Sikorsky returned to his original plan of controlling the helicopter's horizontal flight. The outrigger-mounted auxiliary tail rotors filling in for the initially unworkable cyclic control were mechanically troublesome and the aerodynamic drag was hindering forward flight. The cyclic was a problem that had to be solved.

"He couldn't understand at first why things didn't work," Sergei Sikorsky says. Specifically, the helicopter refused to go in the logical direction. The answer was to set the controls so that the pilot began changing the rotor pitch a quarter-turn ahead of when common sense suggested he should. That's how spinning gyroscopes respond to a force—by moving away at a ninety-degree angle—and a helicopter rotor in motion is a gyroscope of terrific power.

On December 31, 1941, Sikorsky's cyclic finally began working.* At last United Aircraft could see the prototype helicopter that he had promised them in 1938. It was still called the VS-300, but had been rebuilt eighteen times. "Sikorsky's genius was empirical," says Roger Connor, curator of the Smithsonian Institution's vertical-flight collection. "The problem was that he was not a great manufacturer, even when building the flying boats. His aircraft were more crafted than produced—the assembly line was not his thing."

The army-funded successor to the final prototype, the VS-300A, received the formal name R-4, which the British named the Hoverfly.** Now it received all the attention. Only the Focke-Wulf group in Germany had gotten to the point of making routine flights, but the United States planned to go much further. Beginning in 1942, Sikorsky's men had the time to get creative about what the VS-300A and the Hoverfly could do when called upon. Acceleration? It could stop and start faster than a car, reaching speeds of eighty miles per hour. *Life* magazine brought out a photographer who showed Sikorsky tucking the VS-300A into a line of parked cars: "Sikorsky's amphibian helicopter is easier to drive than an auto," read the headline. The men tossed out dummy bombs from the cockpit to see if they could hit anything. They hung a rope ladder from it and did practice rescues. The helicopter landed in a backyard with twelve feet to spare.

Morris flew the R-4 Hoverfly from Connecticut to Ohio in May 1942, which served as official delivery of the finished product. Volunteer aircraft spotters called it in as acting suspiciously; one ob-

---

* Raoul Hafner had worked out many of the details three years before, by incorporating cyclic and collective control into a rotor hub with flapping blades.

** Early military nomenclature for helicopters is difficult to follow because each branch assigned its own numerical designation to a given model such as the Sikorsky R-5. To make it easier for readers to keep track of a given helicopter, I will use a popular name if available (an example is the Chinook) and otherwise will use its most common designation. For instance, only the British version of the R-4 was known as the Hoverfly, but this book will use that popular name for all R-4 helicopters.

server phoned that a farm windmill had just passed overhead. Passing through the Alleghenies, Morris detoured to chase deer across a meadow. Sikorsky was jubilant; *Atlantic Monthly* published an article by him four months later called "The Coming Air Age" that predicted that families would go camping in helicopters with balconies. The helicopter would take ordinary people into remote and roadless places. It was not that Sikorsky figured helicopters would replace cars; he expected them to offer something entirely new.

Now that Sikorsky's R-4 Hoverfly was under control, it was time to see what a helicopter could do for the world.

Chapter Seven
Filling the Sky

The first rescue of a civilian came in April 1944, when a Coast
Guard R-4 Hoverfly came to the aid of a teenager
stranded on a sandbar in New York's Jamaica Bay.
Because the tide was rolling in at the time, the Coast
Guard station at Floyd Bennett Field tallied it as a life
saved. In his office at the Sikorsky plant overlooking
the Housatonic River, Igor Sikorsky began keeping
score of every additional person helicopters plucked
from the icy grip of fate. He wrote congratulatory
notes to helicopter crews he read about in the paper.
"If a man is drowning in the ocean," Sikorsky liked
to say, "an airplane is only good for tossing him a funeral wreath. But a helicopter will bring him home."

Such rescues—and there would be many in war
and peace—made news throughout the helicopter's
first ten years of employment. The machine's amazing versatility suggested to some people that helicopters could find a brisk demand after the war,
comparable to that for upscale automobiles, if
enough manufacturers materialized to satisfy consumers. Hundreds of would-be Igors rose to the

challenge. At least two dozen raised enough capital to go into business, however briefly.

But would enough of America's homes warm up to what one writer in December 1942 ungraciously called "the crackpot's dream"? It was apparently assumed that husbands were fully committed but their wives' allegiance was in question, even though the great majority of articles suggested that flying a helicopter would be as easy as operating an elevator or driving a car. Therefore a number of articles referred to the housewife who would push a button or raise a lever and rise to cruising altitude, then would push or pull something else and zoom horizontally to a destination as if steering a car along an electronic highway. There was blithe talk of landing such machines in backyards, perhaps ones as small as twenty feet across.[1] Charles H. Kaman addressed any doubters by arranging a demonstration for *Life* magazine in 1948, in which he loosed Ann Griffin, "young housewife of Simsbury, Conn.," on a solo flight in his K-190 intermeshing-rotor helicopter, after thirty-six minutes of instruction. Photographs and captions made the point: Anybody could do this thing. Still, poet Phyllis McGinley of the *New Yorker* expressed some doubt in a stanza from her 1943 work "All God's Chillun Got Helicopters": "Still will the Sunday pilots soar / Reckless of holiday disaster, / To meet it as they did before / But somewhat faster."

And had the housewife market been pursued exactly as promised, family disasters would have followed. Rising vertically to cruising altitude, followed by horizontal flight, is no way to fly a helicopter. (The early helicopter pilots already knew this. Particularly in single-engine helicopters, pilots need to plan for engine failure at all times. It means that during a landing approach, and until the last seconds before touching down, the helicopter should be brought down diagonally and with sufficient airspeed to allow an autorotation maneuver if the engine quits.)

Whether the housewife-helicopter was real or a promoter's bubble, the need for rescue helicopters was indisputable.* The most

---

* As of 1944, there had been few rescues via balloon. One was in January 1932 when a Goodyear blimp pulled two flyers out of the Florida Everglades. In March 1944 a blimp picked up two downed flyers from the mouth of the Amazon following the crash of their bomber. But rescue blimps were cumbersome to keep on alert, slow to reach a distant emergency, and unsuitable for use during storms.

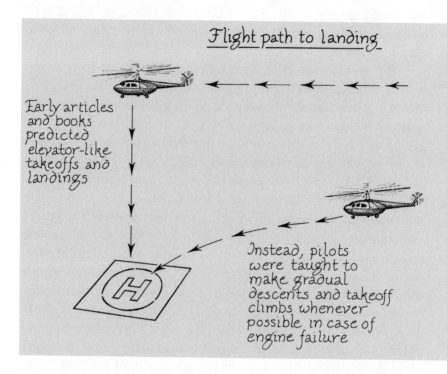

Flight path to landing

Early articles and books predicted elevator-like takeoffs and landings

Instead, pilots were taught to make gradual descents and takeoff climbs whenever possible in case of engine failure

daunting rescue problem arose out of helicopter's rival, the airplane, which weekly plummeted into mountainous or thickly forested regions.

Before helicopters, there was only one way to extract the stranded from a remote area lacking a meadow, gravel bar, plateau, or ledge big enough for an airplane to land. Along with an airplane, it relied on a rig called the Man Harness, devised by All-American Engineering Co. of Wilmington, Delaware. All-American had started in business by producing the means for low-flying airplanes to pick up mailbags at remote post offices. By World War II it was standard practice for a lone commando or secret agent to prepare his own pickup by setting up two poles supporting a large loop of rope; this served as a lift line, which led to a harness worn by the agent, like that of a parachutist. A suitably equipped airplane came in low, trailing a special hook, and snatched the lift line. The agent took flight immediately. The key was a braking winch aboard the airplane that spooled out enough cable in the first seconds to ease acceleration and prevent whiplash. Then the same winch reeled the secret agent into the airplane. The system was so capable that a big enough airplane could snatch out a full-sized Waco transport glider. As with para-chuting, All-American's system needed special training and equip-ment. With some injury cases and in some terrain it was unsuitable.[2]

Meanwhile, Sikorsky's R-4 Hoverfly was making a name for it-self on the home front. The first dramatic news came out of New York's Floyd Bennett Field, where the Coast Guard had stationed helicopters to develop rescue techniques. Early on the morning of January 3, 1944, the destroyer USS *Turner* blew up while anchored in New York's Ambrose Channel. Coast Guard boats took more than fifty badly burned sailors to Sandy Hook, but the men needed blood plasma immediately and other aircraft were grounded by snow, wind, and low clouds. Deciding that cars and boats would take too long, Commander Frank Erickson agreed to give it a try via helicop-ter. Erickson left the field, picked up two cases of plasma on Manhattan's South Ferry, and headed for the beach at Sandy Hook. "The weather was beyond bad," says Sergei Sikorsky, who was sta-tioned at Floyd Bennett Field as a young mechanic. "Erickson felt his way across the harbor by dodging between ships."

Though the R-4 Hoverfly was underpowered, had only a 120-mile range, and was not a top performer in tropical temperatures or at high altitudes, it successfully carried out its assigned tasks in the South Pacific: combat evacuation and expediting the repair of fixed-wing aircraft on island air bases. By September 1944, almost three dozen Hoverfly helicopters were in action. Sikorsky's improved R-6 Hoverfly II helicopter arrived in the South Pacific in the spring of 1945.

Lieutenant Carter Harman made the first combat rescues in April 1944, to assist an expedition into Japanese-held territory by General Orde Wingate's Chindit force. The name was a corruption of *chinthe,* a Burmese term for the winged lions whose stone statues are common to Buddhist temples and monasteries. Wingate chose this symbol for his force, technically the Seventy-seventh Indian Infantry Brigade.

The Chindits' first such expedition, in the spring of 1943, had been done in the conventional World War II manner: troops and pack animals moving across the ground, relying on occasional supply drops by parachute from transport planes. But casualties from hunger and combat were heavy as the Chindit columns crossed rivers and attacked rail lines while being hunted by the Japanese. Prior to the expedition, Wingate had warned the men that anyone too wounded to keep up on foot would be left behind, with a weapon and water. He meant what he said. When on March 6 a Japanese troop truck came upon a Chindit patrol guarding mates who were wiring explosives to a bridge near the Bongyuang railway station, the ensuing gun battle seriously wounded five men from the Chindit No. 5 Column. As Wingate had said he would, Major Bernard Fergusson abandoned all five, including Lieutenant John Kerr.

Only one of five survived capture and captivity, a fact that taxed morale. Upon returning to India at the end of the first campaign, Wingate asked the Americans to use their new helicopters for medical evacuation during his second expedition, which got under way in early 1944. Stationed at the forward air base Aberdeen in Burma as part of the First Air Commando Group, Lieutenant Carter Harman received orders to rescue three wounded commandos from a clearing thirty miles from the base, plus a transport pilot. The early helicopter was not well suited for such marginal conditions, but Harman did it anyway, pulling the men out one at a time and moving them to a riverbank suitable for a light-plane landing strip.

In less famous work, Sikorsky R-4 Hoverflies rode aboard floating aircraft-repair ships, which anchored off island airbases in the South Pacific. These helicopters, along with a small number of R-6 Hoverfly II models, moved parts between ship and shore. As word spread about the rescues in the Burmese hills, the pilots found themselves making frequent combat rescues in the Philippines. One inevitable casualty was engine life, because pilots often had to operate them past the redline limits to complete rescues. The helicopter was bedeviled by a high rate of mishaps, mostly arising from engine failures and pilot error. One helicopter crashed in Burma when it hit a wire.

The Sikorsky R-5 Dragonfly seemed to hold the solution to the Hoverfly's wartime shortcomings. It was three times heavier, which required a leap in engineering. Sikorsky planned a production run of 200 Dragonflies for the British but couldn't meet the original delivery date of 1944. Seeing United Aircraft's production struggles with the R-5, the army asked a refrigerator manufacturer, Kelvinator, to turn out the stopgap R-6 Hoverfly II model. Pilots called it the "re-frigerotor."

The war ended before the longed-for R-5 Dragonfly ever reached the front. Had the United States needed to go ahead with the invasion of Japan, a functional Dragonfly would have been sorely missed, because the army's battle plans had been written with the understanding that it would be available. Following the war, the Dragonfly's civilian version would become the S-51. Along with machines from Bell and Hiller, the S-51 would start to pry open the civilian market.

The success of Bréguet, Focke, and Sikorsky convinced many more inventors and investors to enter the business before the coming boom in civilian aviation passed them by. If Sikorsky could pull off such miraculous rescues with machines barely beyond prototypes, then fully mature helicopters would surely change the face of transportation. Among the helicopter hopefuls was the shipbuilder Higgins Industries, which hired industrial designer Enea Bossi to design a two-seat helicopter for use by sportsmen. Besides that shipbuilder, there was another shipbuilder (Kaiser Industries), a brassiere maker (Newby O. Brantly, who set up Brantly Helicopters), a tire firm (Firestone Tire and Rubber), and a manufacturer of bulldozers and farm tractors (Allis-Chalmers). Most aimed their products for the personal-helicopter market, including the teardrop-styled Safti-Copter, which was billed as safe enough for ordinary people to fly from their driveway.*

Though most efforts never succeeded in selling anything, by 1944 three companies in the United States, besides United Aircraft, were far enough along to suggest that a race to the market was under way. They were Arthur Young and his employer Bell Aircraft Corporation, Stanley Hiller, and Frank Piasecki.

Helicoptrians were aware in 1944 that conventional airplanes had shown no signs (yet) of a mass market on the scale achieved by cars or major appliances; immediately before the war, only one private airplane was sold for every 1,500 cars. And this was despite much effort and expense toward simple, safe airplanes and the

---

* In fact the coaxial rotors of the Safti-Copter were placed too close to each other for safety.

airfields to support them. Curtiss-Wright, for one, had spent millions on research and airfield construction.

The believers looked at the situation this way: Airplanes need timing and coordination to bring in on a runway, but a helicopter could just float back home, needing only the same relaxed degree of attention that a car needs when steered into a parking lot. If lost or approaching a cloudbank, the pilot could land for safety. Since no airport was needed, the personal helicopter could be parked at home and at work. All the nuisance of using an airport would be avoided. A helicopter might look expensive at first, but so had automobiles before Olds and Ford worked their mass-production magic.

It was this set of reasonable expectations that led to one of the most massively wrong predictions in history, comparable to forecasts that atomic energy would provide electricity too cheap to meter, or that mile-high skyscrapers would be connected by an airy web of highways. But it held lasting appeal. While the great majority of experts in the industry gave up the million-copter idea by 1950, magazine covers and children's books nurtured the lovely fable for another ten years. Helicopters were extremely expensive to operate from the first, as revealed by the fact that commuter lines using big transports commonly went broke shortly after federal mail subsidies stopped, or even before.

The three new entries all started with the idea of a small, general-purpose helicopter but soon proceeded down different paths. One of the first projects to get under way was that of Arthur Young.

In 1926 a white, steel-hulled schooner left Germany, bound for America. Reporters were on hand for the *Baden Baden*'s departure because, according to designer Anton Flettner, it would make the trip without sails or engines. It was a "rotor ship," to be propelled by the manner in which the wind moved across two giant rotating cylinders set vertically in the deck fore and aft. The rotors looked like twin smokestacks, though greatly out of proportion. The *Baden Baden* made it to the United States on rotor power in the unexceptional time of thirty-four days. Unmoved by the novelty, the shipping industry encouraged no further experiments.* By that time Flettner was on to other things, such as an airplane that was going to use hor-

---

* The schooner went back to sails and diesel engines, finally sinking in a storm in 1931.

izontal rotor tubes in the place of wings. Later Flettner would drop
that idea also and create successful helicopters for the German
Luftwaffe.

Flettner's fling with the rotor ship caused at least one lasting effect in the United States. It arose out of Flettner's 1928 book about
rotor power, titled *Le Vol Vertical,* which caught the attention of a
philosophically minded college graduate in Radnor, Pennsylvania,
named Arthur Young.

Young was the son of an artist father and a wealthy society
mother, so when he graduated from Princeton with a degree in
mathematics in 1927, he moved back to the family estate and had the
luxury of thinking long and hard about his life's work. At the time
when he saw Flettner's book, Young was looking for a challenging
project on which he could develop some ideas from his college years
about the passage of time and the workings of the human mind. A
photograph in the book made him curious about new types of wind-
mills, and that led him into libraries, where he read everything available about helicopters. The subject interested him because while
proposals went back a century—and in the early years had attracted
more research than fixed-wing airplanes—nobody had built a useful
one yet.

Young gathered pieces from toy stores and hardware stores and
began to build a miniature, one-rotor helicopter with a small propeller mounted on each rotor blade to pull the rotor blade along. It
was a helicogyre, based on the same principle as Louis Brennan's helicopter.

For a full seven years Young stuck with this plan, working in
the family's barn, breaking hundreds of blades and propellers, moving from rubber-band power to a small gasoline engine. Finally he
dropped the propeller-driven rotor blade and shifted to the simpler,
single-rotor design being pursued by Sikorsky in Connecticut. But
rather than attempt to construct a man-carrying version, he confined
himself to electrically powered models that he could control with
wires. Unlike most other helicopter pioneers, he made no attempt to
move quickly to a full-scale prototype.

Young focused on the problem of stability, which was the tendency of the hovering aircraft to move from side to side in a
pendulum-like motion. If not corrected by the pilot, each swing

became more extreme. He thought of two methods, one using the gyroscopic action of a rotating flywheel and another using a weighted bar on the rotor mast. By 1941 he had a model helicopter that could hover well and fly in and out of the barn's doorway.[3]

Larry Bell of Bell Aircraft, a major manufacturer of warplanes, watched Young's demonstration of a flying helicopter model at his Buffalo plant and agreed to hire him to supervise the construction of two full-sized helicopters. Young moved into an empty car dealership in Gardenville, New York. There his team produced its first helicopter, called the Model 30, in six months. It featured a teetering rotor and Young's barbell-like stabilizing bar on the rotor mast, which, as it had in the model, canceled out disturbances when the machine was hovering. Though the Bell plant was at the peak of war production—then adding a thousand employees each week in the rush to build warplanes—Larry Bell took a personal interest in the helicopter project in hopes that it could ease the huge drop in airplane production that was inevitable when the war ended.*

An early movie of the Model 30 shows Bell test pilot Bob Stanley, an expert in airplanes but new to helicopters, flying the ship in a low hover. The helicopter starts leaping and the movie ends with Stanley thrown into a snowbank, scarcely injured. According to the autobiography of Joe Mashman, a chief pilot for Bell, there was an element of showmanship to the early demonstrations for politicians and businessmen. Behind it was a fear that influential visitors might regard a temporary shortcoming in a prototype as a permanent problem for all helicopters. The solution was to shield shortcomings from view. As one example, when an executive wanted to show off the helicopter at the company's main plant in Niagara Falls, twenty miles from the helicopter's home base, workers put the helicopter on a truck and drove it to Niagara Falls, hiding it behind a coal pile at the start of the demonstration and flying it into view at the proper time, as if it had flown all the way from Gardenville. "That shows how little faith we had in the components of our helicopters," Mashman wrote.

---

* Bell felt that there would be plenty of business for all helicopter makers. After the Bell 47 was fully into production Bell received a request for assistance from a young helicopter inventor who was aggressively chasing the same small-helicopter market. Stanley Hiller, Jr., needed rotor blades from Bell's inventory for his competing model. As an act of goodwill, Bell agreed to give Hiller access to his blade manufacturer.

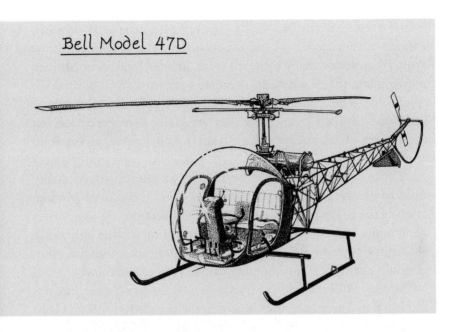

Bell Model 47D

Bell directed Young to build a variant that would be a stream-lined, upholstered four-passenger helicopter, styled like a sedan, suitable for businesses and families in the postwar aviation boom. Young's group assembled a prototype but put most of their effort toward an improved version of their first two prototypes, later called the Model 47—secretly, because Bell had prohibited them from building a third helicopter without permission. In 1946 the Bell Model 47 became the first American helicopter certified for sale to the public. While small, it quickly acquired a reputation for toughness and versatility, acquiring its classic look in the bubble-cabin, lattice-tail Model 47D.

After the Model 47 was on its way Bell continued to experiment with variations from its two-bladed main rotor design. One was the HSL, a two-rotor helicopter intended for hunting submarines by dipping a microphone in the water.[4] Another Bell idea of the fifties, which never made it past a brochure, looked like a giant harmonica with stadium seating inside. The harmonica-helicopter would have flown sideways, giving all passengers a front-row seat out the picture window.

However striking such concept drawings may have been, to television viewers of the fifties there was no other small helicopter in the land other than the Bell 47, because it was the star of *Whirlybirds,* an action-adventure series from the creators of the *I Love Lucy* show. Across 111 episodes, many of which were directed by Robert Altman, the two pilots solved mysteries, cornered criminals, and hoisted the hapless. In "The Human Bomb," a typical episode, pilots P. T. Moore and Chuck Martin think they are going to a charter job but instead meet a bank robber, who hijacks their helicopter to speed his escape.

Those who believe that beauty arises out of functionality would be gratified to know that the New York Museum of Modern Art made the Bell 47 part of its permanent collection.

Despite his success in aircraft, Arthur Young left the business in 1947, saying he wanted to resume his study of human consciousness. He would pursue the "psychopter," the fully liberated mind.

Stanley Hiller lacked the dreaminess of Arthur Young but was fully equal to him in persistence and in the conviction that the helicopter market was ready for a boom. Born in 1924, Hiller got an early start at inventing because his father, grandfather, and great-grandfather were all inventors. By age eight he had salvaged a gasoline engine from an old washing machine and built a go-cart around it. He drove it down the streets of Berkeley, California, until the police stopped him. Later he salvaged a smaller gasoline engine from a model airplane and installed it in a miniature car, which raced around a circular track at speeds up to 100 miles per hour. Making these "Hiller Comets" turned into a major business for him by age sixteen. That year he diversified Hiller Industries into a defense contractor employing 300 people, making high-strength aluminum parts for warplanes. The annual payroll was $300,000. When Hiller was a freshman at the University of California, he got into a dispute with an instructor about the best design for helicopters and his father suggested that Hiller build one to prove his point. Hiller checked into the subject. The fact that none of the helicopters currently making the news in Europe and America used the coaxial approach intrigued him. He might be able to patent valuable ideas since the area was less well developed and it might be safer for home use, since coaxial helicopters don't need tail rotors.

Hiller dropped out of college after one year. He sold off his manufacturing company and began work on a coaxial prototype in

1942 with three employees, working out of a car-repair garage. It was a daring move, because none of the other commercial attempts had gone that route.* Unable to obtain even one aircraft engine because of war-production regulations, Hiller packed his 100-pound scale model in a suitcase and took it on repeated trips to the East Coast until he obtained permission from Washington to buy one ninety-horsepower Franklin engine. On the first runup in the company workshop, the rotors of his XH-44 sucked all the glass out of the skylights. The XH-44 was flying by early 1944. Photographed standing by its side in the shop, the nineteen-year-old industrialist has his jaw firmly set and lacks even the hint of a smile.

Hiller rolled his helicopter over while learning to fly at his parents' house in Berkeley, but repaired the blades and shifted his flying lessons to the Berkeley Memorial Stadium. After mastering it in July 1944, he flew it to the Marina neighborhood of San Francisco to land in a city street. He landed on the rooftop of his Berkeley plant to demonstrate that others could get to work this way. He flew it for reporters in August, meanwhile planning to report for military duty the following month, as he had been drafted. Greatly impressed, the navy awarded the teenager a small development contract and also canceled his draft notice.

As Allied victory loomed, Hiller refined his coaxial helicopter to make it more suitable for daily business use. Hiller thought he had reached a long-term deal in 1944 with shipbuilder industrialist Henry Kaiser to produce it but balked at Kaiser's requirement to move his operations east. Kaiser decided to make cars instead and Hiller started his own company. Unable to interest bankers and security firms in putting up any money, he sold shares in United Helicopters to small investors, most of them civilian war-industry workers who still had some wages saved. Determined to let no opportunity for financing pass untouched, Hiller once called the California Highway Patrol to locate and bring his test pilot back from a honeymoon trip so that a demonstration for investors could be made on time.

---

* One reason is that the two parallel rotors spin in opposite directions and can collide with each other when the helicopter is at cruise speed unless some compromises are made. The blade on one rotor is trying to flap up when the blade on the other is trying to flap down, which will bring them together if the blades are not extremely rigid. To make it fly safely, Hiller's team had to fabricate the first all-metal rotor blades.

Hiller decided to diversify into single-main-rotor helicopters. These possessed better handling qualities than a coaxial, as Sikorsky and Arthur Young had recognized before him. It meant Hiller had to live with a tail rotor, though he felt it was a real hazard for a family copter. He called it the Hiller 360. After the first prototype rolled over and crashed he added a stabilizing device to assist the pilot while hovering, the Rotor-Matic. In the Hiller machine, the pilot didn't move the rotor blades directly. Instead he moved little winglets mounted on the rotor mast at right angles to the two blades, and these winglets moved the blades. It acted like a gyroscope and smoothed the flight so well that while flying over Moffett Field, California, Hiller and his assistant left the cockpit altogether and clambered to the rear, standing by the engine compartment, so a photographer in a nearby airplane could show the open-cabin helicopter flying with nobody at the helm.

By this time, as Hiller worked toward a flyable Model 360, the woods were seemingly filled with aviation entrepreneurs. Reported the *Wall Street Journal* in September 1947, "Scores of energetic craftsmen—most of them young—are at work in garages, back yards, hangars, warehouses and improvised factories, from San Diego to Providence, R.I., competing for one of the big prizes of the machine age: development of a cheap helicopter that can fly easily and safely."*

In April 1944 the owner of Twin Service Stations on Hudson Boulevard East, across the Hudson River from New York, applied for a certificate to operate a helicopter service station. As gas-station proprietor Henry Kiefer saw it, if predictions came true and thousands of people commuted by helicopter, they would need fuel and service work. Further, he predicted, they would be unlikely to find affordable rooftop parking in Manhattan, so he proposed to put a helicopter parking lot on his property, which would allow the pilots to catch a bus into town.

At a stockholders' meeting at the high school football stadium in Redwood City in 1947, Hiller pulled a tarp from the Hiller 360 and invited his investors to choose which way the company should go: the UH-4 coaxial Commuter helicopter originally proposed, or

---

* One entrepreneur named in the article was Gilbert Magill, who raised money for his tandem Rotor-Craft helicopter by selling stock to movie and radio stars.

the new single-main-rotor. They voted for the single-main-rotor model. It lacked the futuristic appearance of the Commuter, but gathered good publicity with rooftop visits, fishing expeditions to remote areas, and a rescue in Yosemite National Park. The publicity would in time bring inquiries from a French dealer who sold two to the French government for military rescues in Indochina. It opened up new civilian uses for farming and construction. Now Hiller needed much more money for factory space to meet the orders coming in. Once again the conventional financial markets refused to invest. By this time the blue-collar investors he had depended on had spent their wartime savings; still, Hiller figured that farmers would come through if coaxed. At age twenty-two, Hiller began construction on the company's first production plant.

Well before the 1948 California State Fair opened, Hiller warmed up the crowd by putting on demonstrations on farms and ranches throughout central California. When the fair opened he flew flight demonstrations at the racetrack. He dusted crops with simulated pesticide and rescued a man who appeared to be drowning in a pond. When each show was over he parked his Hiller 360 among the farm tractors to make the point that his helicopter was the latest in farm implements. The price was $19,995. As spectators arrived from the racetrack to get a better look, stock salesmen invited them to invest. In that manner Hiller sold a million dollars' worth of shares at the fair, enough that production could climb to three per week. Hiller did a victory lap of sorts in early 1949 by flying the last leg of a cross-country trip from his California headquarters to Wall Street. He hovered outside the windows of banks and stockbrokers to demonstrate that, despite their collective cold feet, the Hiller helicopter had arrived.

Stanley Hiller started in business with a clear intention to build and sell small helicopters suited for home and business use. Hiller hewed to that plan until shifting to other industries. Frank Piasecki[5] of Philadelphia started his business with a similar thought, but his audacity gave him an early chance to develop a very different breed of helicopter. Routinely Piasecki rushed into problems so daunting that any reasonable person would have given up early on. He did his own test flying and survived more close calls than had Hiller.

Piasecki's company, the P-V Engineering Forum, assembled its first prototype mostly out of scraps. The airframe came from a wrecked light plane and the landing gear from another plane. Old cars and an outboard motor contributed transmission components. Amazingly, the tiny PV-2 helicopter took flight beginning in April 1943. It was the third successful American helicopter after Sikorsky's VS-300A and XR-4. When Piasecki first flew it he had only fourteen hours as pilot in his logbook, and none of that had been in a helicopter.*

Even so, the public had heard plenty about other helicopter breakthroughs by then and something dramatic was needed to distinguish the Piasecki product. Piasecki obtained critical help from Harry S. Pack, a promoter who talked a Pentagon delegation into putting a PV-2 demonstration on their calendar. He scheduled it for October 12, 1943.

Pack's timing could not have been better. The navy was desperate for a helicopter project. The reason was intense pressure from the Senate Special Committee to Investigate the National Defense Program. Known as the Truman Committee, it had first taken an interest in helicopters in 1942 after a newspaper column criticized the navy's refusal to obtain helicopters, which allegedly could defeat the U-boat menace in the "air gap" portion of the North Atlantic that shore-based airplanes couldn't reach. The committee summoned naval officers and told them that it was simply unbelievable that the Royal Navy needed helicopters urgently but the U.S. Navy didn't. By March 1943, after a long and costly winter battle with U-boats, the British Admiralty had expressed some doubt that their country could continue fighting.

While Allied antisubmarine tactics continued to improve following that crisis, the tone of the message from Truman's committee did not: The navy was to join the helicopter revolution or else. The navy took on a new attitude. Publicly, the navy wanted bigger helicopters than Sikorsky offered; privately, it wanted to avoid purchasing the smaller army-originated models since that would have been to admit the navy had been wrong in avoiding helicopters. That left an opening for a newcomer who a year before would have had no opportunity to enter the field.

---

* By contrast, Igor Sikorsky had more than two decades of flying fixed-wing airplanes before attempting a helicopter flight.

Piasecki's day of reckoning at Washington's National Airport approached. Lacking money for a truck or trailer and unwilling to risk the prototype in flying cross country, Piasecki folded the rotor blades back and towed it there from Philadelphia behind a Pontiac. At the demonstration, a well-tailored Piasecki performed all the things that helicopter pilots are expected to do in flight, moving in all directions without crashing once. The machine looked good as well: Pack had thoughtfully covered it with metal sheeting and plastic, making it look less like an experimental prototype that the P-V Forum lacked the money to replace.

The PV-2 looked so good, in fact, that Piasecki agreed to perform for a newsreel to be titled "An Air Flivver in Every Garage."[6] It was shown as a short before the feature in movie theaters and viewers there could have reasonably surmised that soon any PV-2 owner would be able to roll his machine out of the garage, fold out the rotor blades, toss his golf bag onboard, top off the tank at the neighborhood service station, then flit off to the country club.

In reality, Piasecki's vertical maneuvering at a gas station in Virginia (which was surrounded by obstacles) was dangerously marginal. The PV-2 was so short on lift it couldn't carry golf clubs and a pilot at the same time. The clubs traveled by car, and strategic editing made it appear that the pilot pulled them out of the cabin upon landing at the golf links.

Hokum aside, the P-V Forum's accomplishment in getting the machine to fly at all on such a barren budget, at such an early state of knowledge, was remarkable. Piasecki's team also contributed new ideas that simplified design and reduced vibration. The Navy Bureau of Aeronautics rewarded the Forum with a contract to build the prototype of a very large troop-carrying helicopter. The catch: It insisted on a helicopter that could carry almost a ton. That would have been a major undertaking for any American manufacturer, and particularly for a company whose single prototype hadn't been able to hoist a pilot plus one set of golf clubs. Piasecki sketched and calculated for weeks and proposed to solve the problem with a tandem design: a helicopter with a rotor in the front matched by one in the rear.

In March 1945 Piasecki put his second prototype in the air, officially the PV-3 but dubbed by Piasecki's team the Dogship. The Dogship was a mighty leap beyond the PV-2 in all respects. The 450-

Piasecki HRP-1 "Flying Banana"

horsepower engine sat in the middle of a long, curved fuselage that looked like it had sagged in the middle, with a set of rotor blades topping the fuselage at each end. After a series of hair-raising in-flight hazards* was overcome, this first-generation tandem proved itself sufficiently to justify a production run of large tandem transports for the navy and the army. A Piasecki tandem first lifted a jeep in 1947. In 1949 the marines used the second-generation Piaseckis (the tubular-shaped HUP series) to test ideas about how to get troops onto land during amphibious invasions.

One advantage of the twin-rotor design as pioneered by Piasecki was its tolerance of where the payload went, says Izzy Senderoff of the American Helicopter Museum and a retiree from Piasecki. Early single-main-rotor helicopters were finicky about the placement of weight. The pilots of those had to shift slabs of ballast depending on passengers and cargo.

The military later commissioned a third-generation, banana-shaped Piasecki, the H-21 Flying Banana, which lasted through the 1950s and into the early years of the Vietnam War.[7]

By war's end, the four major American ventures into helicopters—Sikorsky, Hiller, Piasecki, and Bell—had laid out what might be thought as the main tracks of helicopter design. Their layouts are still visible in helicopters built today.

---

* Including an electrical fire, an overheated transmission, and several near crashes, one of which occurred in front of a navy inspection team.

But in helicopters, like much of aviation, there is often room for offshoots that fill a specialized niche. One came from Charles Kaman,[8] a musically gifted teenager who had the opportunity to play professionally for the Tommy Dorsey Band but who had turned it down to study aeronautical engineering. In 1945 Kaman was working in a propeller factory when he decided to set up his own helicopter business. Based out of his mother's garage, Kaman worked out an elegant alternative for controlling the pitch of the rotor blades.* It would be two more years before Kaman began selling his first helicopter. Would returning veterans wait that long to purchase personal transportation? Tens of thousands of well-trained military pilots were streaming out of the service and had wartime savings to spend. War industries were eager to shift into consumer goods en masse. According to the popular press, all pieces were now in place for the personal-aircraft revolution.

* The pilot controlled rotor blades by means of what Kaman called the servo-flap. These movable panels on the trailing edge connected with the pilot's controls via a mechanical linkage down the rotor mast. When these servo-flaps moved up or down, it caused the rotor blade to pitch down or up.

Chapter Eight
The Race

Immediately after the war, with helicopters having performed in very difficult conditions in Burma, China, and the South Pacific, helicopter makers staked their claim on the reinvigorated market for consumer goods. In that spirit, in July 1946 Igor Sikorsky issued himself a challenge. He scheduled a weekday race that would pit one of his $48,000 S-51 helicopters (which he called "my baby, my child") against three old-fashioned means of commuting fifty miles from Bridgeport, Connecticut, to the United Aircraft office in East Hartford. Admittedly, all the participants were employed by the helicopter division or its business associates. One was Igor Sikorsky himself, who would be riding as a passenger in a Lockheed Lodestar airliner.

Any reporter near New York who had ever expressed any interest in helicopters received a limousine ride to LaGuardia Airport the morning of the race. United Aircraft promised to fly them to East Hartford to see the end of the race. It would be expensive, but the industry needed some kind of kick into high gear, and this might help remind a few thousand businessmen that postwar auto traffic was so bad that using a commuter helicopter would extract more productivity out of each executive's workday. It was a chance to craft a new business-like image for the goofy flying machine that had been parodied in a 1931 movie, called *Flying High* and featuring comedian Bert Lahr.[1]

The race began shortly after noon. The S-51 arrived first at the finish line (in just thirty-one minutes), followed by Sikorsky in the airliner, and then the automobile driver (who was so eager to be a contender that she had gotten pulled over for speeding, according to the publicists). The train rider straggled in last. The president of an airline company was riding aboard the S-51. He emerged at the finish line to announce that his company was buying one.

The race flopped as a publicity stunt. The company's publicity department had arranged for a plane to pick up the reporters at LaGuardia Airport, but bad weather had scrubbed the flight and the publicity department forgot to find alternate transportation. The dreadful error had stranded all the New York reporters miles from the action. According to *American Helicopter* magazine, "The result was a bad-humored New York press that except for a very few exceptions refused to even carry the story of the race." The reporters were not made happier when they heard that reporters from small New England papers had gotten to the finish line without trouble.

Sikorsky S-51

Daring rescue by helicopter was a more reliable route to press coverage. Many of the exploits came out of New York. When a workman slipped and fell onto a rooftop perch at the Cathedral of St. John the Divine in June 1951, a police helicopter came and got him. When a fisherman in Jamaica Bay, New York, jumped from his boat to help a man he thought was drowning, the speedboat he left went berserk and began circling at top speed through a flotilla of boats anchored or cruising in the bay. Boats dodged until the NYPD's helicopter appeared. It matched speeds and allowed a detective to drop in. When notified that ex-K9 dogs left over from the war had gotten loose and taken up residence on Sandy Hook, posing a deadly menace to any boater who landed there, snipers went after them from a helicopter. Crews in Oregon discovered that they could rid ranches of coyotes by lowering ropes to them; for some reason the coyotes gripped the ropes in their teeth, hanging on like grim death even when the helicopter rose far into the air, even when a crewman dropped his end of the rope from hundreds of feet up.

In February 1947 a helicopter landed near a shack in the Adirondacks. For thirty-three years it had been the hideaway of long-bearded hermit Noah Rondeau.* But no more: a helicopter flew him and a pile of bows, arrows, fur garments, and other handmade gear to a sporting goods show at Grand Central Palace in New York City. "Up to now I felt the air was only for birds, but now I feel aviation is here to stay," he told reporters.

The most remarkable and sustained publicity about the rotary-winged revolution came out of Texas, where, in the spring of 1948, a young and lanky congressman had a problem. Representative Lyndon Baines Johnson had let the filing deadline pass so he couldn't run for reelection to his seat in the House of Representatives. All his hopes had been placed on winning Texas's open seat for the U.S. Senate. But in late May, while he should have been campaigning hard to win the Democratic primary, he had been compelled to check into the Mayo Clinic in Rochester, Minnesota, to undergo kidney-stone surgery. He was trapped there for a week.

The man he would have to beat in the primary election scheduled for July 24 was the immensely popular former governor of

---

* According to a publicist.

Texas, Coke Stevenson. A good case could be made that Stevenson, a rancher, was the embodiment of everything that made Texans feel good about their state. Stevenson had run for office in Texas twelve times in his life, and had won twelve times. While LBJ was known chiefly to his Austin-area district as an FDR liberal, Stevenson was famous statewide and was conservative to the roots of his hair. "Conservative" accurately described the majority of Texans.

Once Johnson got out of the hospital, his campaign worked hard to muster a small and noisy throng to hear what would be the first radio speech of his resumed race for the Senate. His return to the hustings was received with no more newspaper attention than his first speech announcing that he was a candidate, made in Austin's Wooldridge Park. The Senate race was already crowded with aspirants like Johnson, all deeply in the shadow of Stevenson.

Johnson was looking at the very real prospect of losing the primary and heading back home to Austin, where he would manage the radio station KTBC and fulminate about how he might have become a major political figure one day, perhaps, had he done something to steer his fate. Johnson knew he needed to cover his vast state with a hard-hitting conservative message. The message itself was easily decided: attack Coke Stevenson as a flip-flopper because he would not take an explicit stand on the Taft-Hartley Act, which when passed would cut back on the power of labor unions. But the message was of no use unless Johnson could get to the little towns that Stevenson, being only human, would miss while campaigning. There were hundreds of such forsaken and forgotten towns, connected by many thousands of miles of dusty farm roads: more towns than anyone had dreamed of visiting over the space of a few weeks. Johnson would get to nearly 400 of them.

Johnson had to turn things around, and when it comes to turning things around, what better than a rotary-winged aircraft? Warren Woodward was the first to suggest helicopter campaigning to Johnson. After seeing a demonstration of helicopter flight in Washington during the spring, Johnson agreed that this looked like his single chance to be noticed in such a short time. Indeed, very few residents of rural Texas had ever seen a helicopter. There were only a handful of civilian helicopters in the world in June 1948 certified to carry passengers and powerful enough to climb out of town squares in a scorching Texas summer. Fortunately the Sikorsky Division of United Aircraft happened to have such a ship on hand, a used but serviceable S-51 model that resembled the wartime Dragonfly, but with better performance.

"The campaign said they were paying for it," says Harry Nachlin, assigned as the crew chief, "but what they paid wouldn't have covered even my salary." Whenever asked about the expense, Johnson answered that a hundred patients at a veterans' hospital (supposedly a committee, the "Dallas Veterans for Johnson") had contributed five dollars apiece. It was laughable to anyone in the industry who knew the daily costs—$250 per day—but it sounded plausible and it papered over the fact that Johnson was using enormously more cash in the primary than campaign laws allowed, which was ten thousand dollars in direct spending.

The ship LBJ would be using had been one of the first ten S-51s off the production line. The fact that it was available for lease on short notice is an indication of how dependent the helicopter industry of 1948 remained on military purchases.*

The first owner had been the bus line Greyhound, which had begun planning to branch into helicopter service in 1943 and, in time, to offer helicopter connections to a thousand cities and towns. Said company president C. E. Wickman, "If our application is approved we plan to make air travel available to the millions who live scores or hundreds of miles from the large airports that present airliners must use." The plan was to begin the heli–bus line[2] out of the Great Lakes region since the president of that division, Manferd Burleigh, was the most enthusiastic helicoptrian among the bosses of Greyhound buses.

As with many such projects, the beginning of the Greyhound experiment was publicized but the end was not. The Civil Aeronautics Board regulated all air transportation. At first the CAB granted permission for a limited run (between Detroit's bus station and airport, and out to Bay City) but later withdrew permission after Congress sounded alarms about a looming bus-helicopter monopoly.

---

* This pattern continued through the Vietnam War. Military needs accounted four-fifths of helicopter sales by Sikorsky and Bell as of 1970.

Meanwhile, the idea was unraveling on its own. The two S-51 helicopters that Greyhound bought to try the idea held only three passengers but cost the equivalent today of a thousand dollars an hour to
operate. In the end Greyhound accomplished little more than to give
free rides to opinion-makers across the upper Midwest. The company also sold five-dollar rides to tourists in Florida and hauled mail
on an experimental basis. "When all the bills started showing up at
Greyhound," says Nachlin, "they called up [the division president]
and said, 'Sell 'em back.'" The Sikorsky Division of United Aircraft
repurchased both helicopters, keeping one for promotional purposes
and the other for testing.

So it happened that a two-ton ex-Greyhound helicopter was
available for use on short notice. It was not the first time a candidate
traveled by helicopter,* but it would be the first time helicopters
made a difference in the outcome. On June 10, 1948, at the Stratford
plant of United Aircraft, two Sikorsky employees climbed aboard the
helicopter registered N92805, soon to enter political lore as the
Johnson City Windmill. One was a former B-25 bomber pilot named
James Chudars, and the other was the crew chief, Harry Nachlin.
Nachlin knew the ship well, having been the crew chief for the Great
Lakes Greyhound experiment.

Johnson's air assault on rural Texas began just before nine in the
morning on June 15, 1948, when he dropped into a softball field in
Terrell, thirty miles east of Dallas. Five hundred people had gathered. Johnson played the part of the indignant conservative, waving
his arms in anger about Truman's civil rights program (a "fraud and a
sham" that he would fight) and any federal attempt to control
schools. Then he ascended into the heavens and headed off to seven
more stops for the day.

Within a week the Johnson team had figured out how to
cover the most towns from dawn to dusk, how to refuel, and how to
stir up a crowd. Running a helicopter campaign required many more
staffers than the usual ground attack, but it was worth it. After
one week, polls showed that the gap between Stevenson and Johnson

* In October 1946, Senator Alexander Smith was the first candidate ever to use a helicopter. He
chartered an S-51 from Helicopter Air Transport.

had narrowed from thirty-six percentage points to ten, a stunning development. Suddenly, the man who had barely been able to turn out old friends to hear his studio speech was looking out over week-day crowds that exceeded a town's entire population.[3]

For any given town or city it worked like this: Days ahead of the visit, college students and local Democrats cruised the streets to pin posters and distribute handbills. Local officials identified a land-ing spot, trying to mesh the directives from Johnson (put me in the center of town) and from Chudars (give me room to land without hitting anything). The day of the Johnson visit, one or two cars with a roof-mounted bullhorn raced into town from the previous engage-ment, advising all to "come to the speaking."

"Of course, mostly they came to see the helicopter," political operative Tommy Corcoran told LBJ biographer Merle Miller. "They'd never seen one before. Christ, it was brilliant as hell."

As Johnson waxed indignant about President Truman, his aides were blazing down the roads to reach the next town, hitting speeds of ninety miles per hour to beat the helicopter, which could fly a bit faster and in a straight line. One advance man was Sam Plyler, who had orders to be on the scene at the moment of landing to guard the tail rotor and prevent people from blundering into it.

When the Sikorsky arrived, it took up an orbit near the landing spot proposed by LBJ's staff. As Johnson bellowed over a public-address system bolted to the landing gear delivering the news that he would soon be speaking, Chudars scrutinized the spot. If he didn't like it he headed for a better one, usually a ball field. Johnson raged in the early days but learned to live with it.

"The only guy that could get Johnson to do what he wanted was the pilot," Nachlin says. "In fact LBJ would introduce him that way: 'That's Jim Chudars, that's the only guy who can say no to me.' He couldn't care less what LBJ thought."

The helicopter was painted with Johnson's name in huge block letters. As it approached the ground, Johnson threw his twenty-five-dollar Stetson into the throng, getting the hat back later by having his staff chase it down or by paying a dollar to the child who brought it up. As the rotor stopped he climbed atop a wood crate, picked up a microphone, and gave a short speech naming local projects and prominent citizens. Then he surged into the crowd, grabbing hands and yanking each person past him, then reaching for more. Everything had to be done on the run so he could get to all the towns on the daily list. Some late arrivals to the speeches didn't believe the helicopter could fly at all and demanded that campaign workers show them the flatbed truck on which it must have arrived.

In a biography of his brother LBJ, Sam Houston Johnson recalled that a farmer in Kickapoo, Texas, had only one thing to say after the speech. The farmer turned to his wife and said, "If he can keep that damn thing from chopping his head off, he might make a good senator."

"When it was getting ready to leave, people would get back but they'd never get back far enough," Nachlin says. "The pilot would wave 'em back and when the dust started to fly they'd wish they'd gotten back like he said."

Johnson used the Windmill to campaign between towns whenever the opportunity arose, landing near construction crews to shake hands or using the helicopter-mounted bullhorn to shout at farmhands from the air: this happened whenever Johnson saw "more than two people and a big dog," according to Chudars. If the helicopter had to land because of a storm, Johnson had the helicopter touch down in a farmyard and then barged right into the house to ask for shelter and a vote. On a single day in early July Johnson made two dozen public appearances. The helicopter took a break only on Sundays, for repair and inspection. LBJ slept no more than three hours per night and drove everyone around him without mercy, berating hotel workers as well as longtime staffers.

Over three weeks the Sikorsky flew Johnson for a total of one hundred hours, then headed back to Connecticut for a major overhaul.* No other Sikorskys were available, but Johnson was able to continue flying through the primary because a competing model, the Bell 47B, was available.

The Bell was smaller and more maneuverable and so was easier to slip into tight spaces. On one occasion the pilot, Joe Mashman, set it down on the roof of a gas station. But the 47B had less power and many takeoffs were done on the ragged edge. "I hadn't met the

---

* Once repaired and repainted, the Johnson City Windmill (N92805) continued to fly commercially into the 1970s, mostly for utility-line construction. As of 2006 it was preserved at Fort Rucker, Alabama.

congressman up until then, and I was dismayed to find out how much he weighed, his size," Mashman told LBJ biographer Merle Miller. "Coupled with my size and weight—I was up at about 185—I knew that performance was going to be very marginal."

If the pilot needed the machine as light as possible for takeoff, Johnson announced to the crowd he would join Mashman later outside of town at a clear landing field, but teased them with suspense first: would the brave pilot be able to climb over the oak trees to the west, or the telephone wire over Main Street? And the risk was real, given the difficult flying conditions. This was a time when civilian helicopters were suffering one accident every 740 hours (a hazard rate twenty-five times higher than today). With a few close calls, one involving an abrupt plunge to a street when the helicopter got caught in its own downwash, Mashman's Bell Model 47B flew LBJ another 7,000 miles on the second leg of his primary campaign.

Though the helicopter was best suited to reaching a rural electorate, Johnson used it occasionally to drop in on suburban shopping centers in Houston, Dallas, and Fort Worth. By August 1948 the focus of the primary campaign was on the big-city vote, where helicopters offered no major advantage in reaching voters. Johnson switched to traditional campaigning by car. But the helicopter had made its mark; on July 24 Johnson had won enough votes to reach a runoff primary the following month, where he went on to defeat Stevenson, followed by rather plausible charges of ballot-box stuffing.

"It was a case of getting close to them," said Chudars. "Without the helicopter, he never would have seen all these people."

LBJ had been taking more risks than he knew. The day the Sikorsky departed the campaign, while over pine forests near the Arkansas-Texas border, the engine quit in midair. Chudars was able to restart it on the way down. Examination showed that the chamois cloth that Nachlin had been using to strain grit out of the gasoline on fueling had distintegrated and left a linty residue in the gas. This lint had temporarily strangled the engine's fuel supply. The problem required a complete disassembly of the engine back at the factory. It also revealed that dust from the many off-airport landings had scored the cylinder walls.

"We'd been flying LBJ the day before we left," Nachlin says of the engine failure. "Had [the engine failure] happened on takeoff, the history books might have been written differently."

The Johnson City Windmill was a boon for Sikorsky and Bell, and other politicians would seek out helicopters for the 1950 election, but such work was too occasional to build a business around. In fact, no single job seemed to be steady enough. The helicopter business was looking like a stream of itinerant odd jobs, as described in Alexander Klemin's 1948 book for teenagers, *The Helicopter Adventure.* Whatever work paid cash got done.

A living example of the itinerant helicopter trade was Bob Trimble of California, who went from job to job in hobo fashion, hauling his Bell 47 on a trailer behind a car. That allowed him to chase work that ranged from firefighting among the lodgepole pines to uranium prospecting in the Grand Canyon to wilderness surveying for the government to hauling lumber for miners' cabins. Other helicopters were doing the same thing all over North America: hauling 200 tons of equipment for building a dam high in the Canadian Rockies, scouting for oil in the marshes of Louisiana, and dragging barges in the Arctic.

The day-to-day uses that emerged during this time of experimentation were much along the line predicted two decades earlier by Nicolas Florine, who had built the first working tandem helicopter. Florine suggested that helicopters would be most valuable for spotting schools of fish, aerial photography, traveling between downtown areas and an airport, tourism, artillery observation, antisubmarine warfare, and border patrol.

One helicoptrian well trained by the war was Knute Flint, a pilot for the air force who made rescues in China in 1945 with a Sikorsky R-6 Hoverfly II. Flint and a few backers created the Armstrong-Flint Helicopter Company and bought two Bell 47Bs that served as the seeds of an international helicopter charter business. Flint's helicopters flew components of a church into the Grand Canyon, filmed the movie *Johnny Belinda,* and sprayed crops. In 1953 he began flying supplies into New Guinea for an oil company that had found it impossible to carry out its explorations with parachute-dropped supplies or airplanes. In just half a day, Armstrong-Flint crews would hack openings in the rain forest that were large enough to fit a helicopter, then build a landing platform atop the pile of wood. In Egypt, when an oil company found it impossible to operate in areas infested by mines laid during World War II, Flint brought in bomb experts who reached out from a helicopter with a mine detector to find a space big enough to put their feet upon; once on the ground they cleared a landing zone for the helicopter.

Santa-hauling provided a steady if seasonal business to Armstrong-Flint. Helicopter Air Transport of Camden, New Jersey, had begun delivering Santa Clauses to department stores in the Christmas season of 1946, reaching six department stores that year, using parking lots and improvised rooftop pads. Major department stores had been looking for a way to use helicopters year-round to keep their downtown stores going as shoppers moved to the suburbs. One solution drawn up by the Kerr's chain of Oklahoma City, and presented to a 1943 helicopter convention, proposed that helicopters pick up shoppers and bring them to a suburban Park-O-Port with a helipad and plenty of parking. Buses would be available at the Park-O-Port to run the shoppers downtown, until the stores had a chance to build their own rooftop heliports. It sounded workable, but there was little to be done about it during the war since consumer goods and private aircraft were all in short supply.

Enter Adam's Helicopter Show, a traveling display featuring an Aeronautical Products Model A3 personal helicopter. The helicopter inserted itself into the floors of many large department stores to convince Americans that a transportation revolution would follow shortly after the end of war rationing. "Enthusiastic plans are being made everywhere for using the machines, which can be flown from backyards, roofs, or any small cleared space," according to an article in the *Wall Street Journal* in 1943.

For the 1946 production of its annual fashion show, the *New York Times*'s fashion writers stitched helicopters into the plot: a millionaire in a penthouse apartment is startled by the descent of a Bendix helicopter onto his rooftop patio. Fashion model Nan Green steps out of the machine, starts up a whirlwind romance, and in the second act goes on to model the latest in women's flying costumes "for college and business girls who are studying to pilot their own planes." Similarly inspired, Filene's of Boston sold a fashion it called the "helicopter dress of tomorrow."

After the war, the G. Fox store in New Haven, Connecticut, used four helicopters to move goods from the central store out to suburban stores and customers across the state. But as with so many helicopter news items, each announcing the revolution, it was a short-term promotion, done in this case to mark the company's hundredth anniversary.

But helicopters still had work to do even as the retail focus moved from downtown stores to shopping centers and malls. They hovered over grand-opening crowds and dropped coupons, sourball candy, and numbered Ping-Pong balls, some of which earned a generous prize. Operators began having second thoughts about such promotions by the late 1950s. One was James Gavin's charter company, Mercury Helicopter Service. Gavin was a leading figure among early helicopter users. By 1962 he refused all jobs that might bring his helicopters in close proximity to crowds of excited children. The larger prizes were inducing adults to take shortcuts by scrambling across parked cars. And hard candy flung from a helicopter could catch children and their parents in a pelting hail.

Live news video from a helicopter, and also live TV coverage of breaking news in general, has roots in KTLA, the first commercial television station on the West Coast. The man behind KTLA was Klaus Landsberg, an electronics technician who emigrated from Germany in 1937 rather than join in the war effort there. Landsberg got a job helping to prepare a demonstration of television at the New York World's Fair. Later he took the job of opening up a station in Los Angeles for Paramount Pictures. It operated experimentally beginning in 1941, and went commercial in January 1947 with Bob Hope serving as emcee for the variety show. The guide for programming was simple: whatever the imperious, hard-driving Landsberg thought the public would like. Landsberg was the first to put Marilyn Monroe on television. Other stars included Bob Clampett and his puppet Beany, Ina Ray Hutton and her All-Girl Band, Harry Owens and his Royal Hawaiians, and a bandleader named Lawrence Welk, whom Landsberg found at a ballroom in Ocean Park.*

* Landsberg also gave the green light to "Musical Adventure with Korla Pandit," featuring an Indian organist-mystic in a turban who said nothing at all in 900 episodes, only playing the organ while clouds scrolled by on a screen behind him. In another program, a Latin lover by the name of Renzo Cesana sat at a tiny table and recited romantic monologues to women viewers.

In 1949 Landsberg set aside all his programming, corny as well as groundbreaking, after a three-year-old girl named Kathy Fiscus fell into a well in San Marino. As rescuers mobilized a frantic rescue operation, Landsberg sent a KTLA crew to the scene and broadcast live and continuously for nearly twenty-eight hours, bringing ordinary routines to a halt across Southern California as viewers agonized over the outcome. The public engaged so strongly with the newsmen that after the county sheriff was told the girl had been found dead, the sheriff asked newscaster Bill Welsh to be the first to inform the Fiscus parents. "The Kathy Fiscus story was the turning point for television in Los Angeles," Stan Chambers of KTLA said later. "Until then, television was just a plaything."

Landsberg began using helicopters in 1952, in the process of setting up a remote transmitter and relay antennas for the broadcast of an atomic explosion in the Nevada desert. After early experiments back in Los Angeles that involved putting a television camera aboard a helicopter, climbing to an altitude of fifty feet and hanging a heavy cable down to the ground, KTLA and National Helicopter Service joined to create the revolutionary Telecopter in 1958. This camera-equipped Bell 47 could transmit a picture without being tethered to the ground, by relaying the camera signal to a receiving antenna on Mount Wilson. When the Baldwin Hills reservoir began leaking on December 15, 1963, pilot Don Sides was on hand for KTLA in the Telecopter and was broadcasting when the dam broke and sent a fifty-foot wall of water along Cloverdale Avenue, wiping out five dozen houses. More than any other single event, says Dick Hart of National Helicopter, Baldwin Hills was the break that the Telecopter needed; until then no other station had been interested in the expensive technology. The Telecopter got close enough to rioting in Watts in August 1965 to receive gunfire.

What about all the personal helicopters promised in one of *Popular Mechanics'* most famous illustrations? On the cover of the February 1951 issue, a homeowner was shown stowing his personal helicopter in a garage. Estimates of that market ran to a million helicopters or more.* The idea of tens of thousands traveling into a city

---

* Common estimates published during the war predicted family-helicopter prices between $1,500 and $2,500. The reasoning was that existing military helicopters could be easily converted for private use, and then produced in mass quantity.

by personal helicopter may seem daffy now, but during the first years of helicopter marketing, before the construction of vast networks of federally backed interstate highways, automobile travel was so frustrating and dangerous in densely populated areas—not just on city streets but on the two-lane highways between cities—that almost anything looked better than cars.

Part of the blame for the boom in street traffic can be laid upon the Olds Motor Works of Detroit, which in 1901 figured out how to make and sell affordable cars. It was a shift of enormous significance because cars were a European invention and the custom abroad had been to offer handmade cars priced for the wealthy. Olds simplified other carmakers' designs and came up with a plain-vanilla model, to be built entirely with parts suited for mass production. By buying these parts a thousand at a time, the company drove down retail prices and showed the way for other entrepreneurs. The number of cars registered in the United States went up by a factor of 150 in the years between 1895 and 1905. From 1910 to 1930, car registrations went from a half million to 22 million.

The highways weren't ready. In 1928, yearly automobile deaths hit 28,000, then jumped another 3,000 the following year. Typical highways were only two lanes wide and jammed with cars turning at driveways to reach stores and restaurants. U.S. Highway Number 1, running down the Eastern Seaboard, was the worst, with more than sixty businesses per mile across one fifty-mile stretch in the 1930s. Americans knew what they were missing, because the World's Fair of 1938–1939, and many magazine articles, promised that superhighways were going to release drivers from all such worries.

By 1940 many cars were capable of cruising at seventy-five miles per hour. Drivers were tempted to pass any slower vehicle at the first opportunity, even if it required driving with maniacal abandon. The death toll hit 45,000 in 1941, then plunged when wartime measures ordered the rationing of gasoline and tires and suspended car manufacturing. But when rationing ended, the scourge was back. President Harry Truman called yearly meetings to sound the alarm. At the President's Highway Safety Conference of 1946, he denounced the "morons and crazy people" who had brought the traffic-death rate up by 45 percent from the previous year. More

Americans had been killed in traffic, he said, than in combat over all the wars the country had engaged in, including the French and Indian War.

Even with Truman sounding the alarm, and even with contractors and automobile dealers firmly behind the federal plan for more and wider highways, only the states took action. As of 1950 the only four-lane, high-speed, divided highways available in the United States were isolated stretches in New York, Connecticut, California, and Pennsylvania. The first was the Bronx River Parkway; the most ambitious was the Pennsylvania Turnpike.

With highways so congested and dangerous, it looked like the opportunity for personal helicopters was still open. Hiller had tried the commuter-helicopter market in the late 1940s, and had given it up in favor of helicopters for business and the military, but to some helicoptrians his departure was good news. It meant that opportunity was still knocking during the late 1950s.

One of the last entrepreneurs to take a chance on the retail helicopter market was Rudy J. Enstrom, a mine-equipment mechanic living at Crystal Falls in Michigan's Upper Peninsula. In 1942 Enstrom started building a helicopter in his father's sawmill, relying on trial, error, information from the library, and a course he had taken on diesel engines. The prospect of a homemade helicopter terrified his mother, but he continued. After four years of striving, he got one of his aircraft to hover. He worked over the final prototype in the basement of his house. For each test he dragged it out through extra-wide doors and hauled it via trailer to a gravel pit, where he tested it in the twilight, relying on his wife, Edith, or a neighbor for help.[4]

Enstrom's test flights had to be conducted in a low-profile manner since he had no experimental-aircraft paperwork and no pilot's license. He constructed five homemade helicopters through 1958, at which point civic leaders of Menominee, Michigan, decided an aircraft industry in that town might bring the local economy through a crisis caused by the closing of a large iron mine. Enstrom was more than willing: he worked at the mine and the shutdown had thrown him out of work also. Seven trying years of trials followed, during which six thousand Michiganders invested in R. J. Enstrom

Corporation to get the three-seat F-28 model certified for sale.*
Some tried to invest their life savings, and were gently dissuaded.
One farmer literally pulled the cash out of a woodpile. "We've stuck
our chin way out for Enstrom. I hope they sell 'em like Fords," said
stockholder (and mayor) John Reindl.

Meanwhile, just as Enstrom and other latecomers, including
Cessna Aircraft,[5] were trying to gain a foothold in the personal-
helicopter business, a fleet of bulldozers and concrete mixers was
making the helicopter less important as an alternative to automobile
traffic problems. President Dwight Eisenhower had helped break the
political stalemate that had held funding back. With passage of the
Federal-Aid Highway Act in 1956, the annual spending for divided
highways went up fourfold. "No one mentions it out loud, but there
is apprehension in the industry over the Federal Government's proposed highway program," wrote *Aviation Week* regarding the helicopter's sales prospects that year. Railways and helicopter airlines had
once hoped to profit from short trips up to 250 miles but now saw
that prospect vanishing. "There is no doubt about it," the reporter
concluded. "The turnpike building program is a major factor in this
picture." If there ever were any prospect of helicopters in garages
across America, the spread of high-speed highways strangled it by
1960.

But all was not lost: The military had started helicopters off in
the first place and, if persuaded, had the money to keep them going.
A few atom bombs would do the persuading.

* In early 1961 an engineering team headed by Alb Ballauer had to scrap most of Enstrom's early work
and start a new model. A second crisis came the following year when a prototype crashed and killed
the test pilot.

Chapter Nine
The Tyrant

The B-29 was supposed to take dead aim at the battleship *Nevada* when it dropped the first atomic bomb for the 1946 tests at Bikini Atoll, but something went wrong and the falling bomb was off almost a half mile. It went for the USS *Gilliam* instead.

How were amphibious forces supposed to land and do their job in a nuclear war? This question highlighted a difficulty the army had recognized before the atomic bombs fell, called the "tyranny of terrain." While the air force had dominion over the postwar skies and the navy ruled the seas, the army still slogged across the landscape, where any number of ruts, ravines, or rivers would slow progress and expose its flanks to future enemies. Even at this early date, such problems suggested that helicopters might play a major role in wars to come.

When commissioned in August 1944, the 4,200-ton attack transport *Gilliam* was the first of a new class of attack transports. Two months later it raised steam and departed from San Francisco with a load of 750 troops bound for New Guinea. It served with distinction in the South Pacific, moving manpower for amphibious landings on Japanese-held islands. In December 1944, while on the way to take on soldiers at Leyte Gulf, it helped fight off a wave of kamikaze and torpedo airplanes. It survived another wave of kamikaze airplanes off Okinawa the following April. Okinawa would be its last combat engagement, though it continued carrying troops around the Pacific for another ten months. In February 1946 the voyages of the *Gilliam* came to an end when it anchored in the lagoon at Bikini Atoll. The ship took its place as part of a ninety-five-ship target fleet for Operation Crossroads' first blast. The twenty-kiloton bomb arrived from the B-29 bomber called *Dave's Dream* on July 1, exploding 518 feet above the water and 150 feet off to the side.

Test Able completely destroyed the *Gilliam*. The concussion sheared the decks from the hull and packed them like pancakes into the ship's bottom. The ship sank in seventy-nine seconds. Altogether, the bow of the ship looked like it had been caught between a giant hammer and an anvil. Another attack transport, *Carlisle,* lay a third of a mile farther from the zero point of the blast; it sank also, its main deck dished downward. And the underwater blast later in July, Test Baker, sank four times as much tonnage.

Equally damaged was General Roy S. Geiger's confidence in the future of conventional amphibious landings. Geiger, a veteran of several hard-fought landings including the one at Okinawa, was on hand to observe the atomic tests for the Marine Corps. Geiger immediately sent a memorandum to the commandant, arguing that the marines could no longer depend on massed fleets of ships delivering troops to storm a disputed beach. Such a huge concentration of shipping could be wiped out by any nuclear-capable enemy. Helicopters, the marines decided, offered the solution. Troop-carrying helicopters would lift off from aircraft carriers, which would be so well dispersed across coastal waters that some would survive nuclear attack. The marines of Helicopter Squadron HMX-1 got the assignment to practice the new style of heliborne amphibious landings during maneuvers in May 1948, using Sikorsky R-5 Dragonfly helicopters.

The Soviet Union acknowledged the same problem, but only after the start of the Korean War. In October 1951 Joseph Stalin cut orders to develop helicopters for war with the West. These projects led to the early transports Yak-24 and Mi-4. Their projected role would be airborne assault behind enemy lines, in Russian *vertoletnyi desant*.*

Still, helicopters for assaulting anything were hard to come by in the early years. This was made clear on a Sunday afternoon in March 1950 when Company A, Nineteenth Infantry Battalion, New York Marine Corps Reserve, needed to demonstrate its skills to marine inspector Captain Richard M. Cook. When the company needed to move out from its training ship at 136th Street to Passaic across the Hudson, it didn't have marine helicopters to make the trip and in the end the bus it hired didn't arrive either. So the troops crammed into sixteen private cars and a jeep and drove across the George Washington Bridge, paying tolls all the way. The inspector tagged along. He watched the marines dismount in New Jersey and then spread out to occupy the objective, a park, on schedule. Captain Cook told a reporter that the battalion commander had hoped to survey the maneuvers from on high, but had been unable to come up with payment for the fee, which was sixty dollars per hour.

By March 1950 such stubbornly high costs, and such customer penny-pinching, had put the helicopter industry in serious trouble. It was entering its third straight year of decline. While sales in 1947 had made it appear that the industry was off to a good start, it was a cruel illusion: the following year suggested that the aviation boom was over. At Bell, helicopter sales fell by more than half. By June 1950, the total of helicopters shipped was a quarter of what the company expected. The company that manufactured the Bendix Whirlaway—the one featured in the *New York Times* fashion show of 1946—went out of business, unable to sell enough machines either to the military or to the civilian market. Aeronautical Products—whose product was featured in the Adam's Helicopter Show department-store displays of 1943—scrapped its only model in 1946. Airplane makers weren't doing any better. From a high of 30,000 units sold in 1946, airplane sales dropped by 50 percent in 1947 and again in 1948.

Still, there were signs that if the industry could hang on long enough, the military might warm to helicopters' potential. Even with the underpowered transport helicopters available, war games on

---

* In the end, Soviet helicopters had a greater effect on Soviet civilian history than its military, because of the role of heavy-lift helicopters (such as the Mi-6 and Mi-26) in opening up mineral-rich areas.

the East Coast in 1949 had convinced the marines that the helicopter would revolutionize landings on enemy territory.

The crisis among American helicopter makers ended in June 1950, when North Korean forces invaded South Korea with assistance from the Soviets and later China. Until August, the North Koreans looked likely to drive American and then U.N. forces entirely off the southern tip of the peninsula. A comparatively few helicopters helped stem the tide while reinforcements mustered. Their contributions were modest at first but got much attention in the press and among commanders. Said one marine commander during the amphibious landing at Inchon, helicopter procurement should trump the purchase of all other weapons.

The army air forces had flown rescue missions during World War II, so it made sense that helicopter missions in Korea would do about the same thing and use much the same lightweight equipment. This was initially the Sikorsky R-5 Dragonfly, a model whose production delays had kept it out of World War II. Comparable lightweight Bell and Hiller helicopters for evacuation and battlefield observation followed soon after. Using such small helicopters, the army, air force, and marines would carry out 30,000 medical evacuation and rescue missions during the three-year war.

The tactical situation in the early weeks was well suited for contributions by small, short-range helicopters. The Allies had begun their counterattack by occupying small beachheads with amphibious forces. The distance to cover during helicopter rescue sorties was so short that even if a helicopter took antiaircraft fire and started leaking engine oil, the pilot could usually make it back to friendly lines before the engine seized up. So the need in Korea seemed to match the need in the South Pacific during World War II. Had it stayed that way, helicopters would have been no more than a footnote in the history of the Korean police action. But opportunities for a new generation of helicopters opened after the Allies freed Seoul and after the North Koreans retreated north across the present demilitarized zone. Distances were now greater and the northern terrain was hillier, making it more difficult for attacking Allied troops to take the high ground in the conventional manner. Helicopters could solve that problem—if they had much more range and lifting capacity than the models from the 1940s could offer.

At the outset of the conflict, the army had a grand total of fifty-six helicopters, all of them small. It wasn't entirely the army's decision: after a 1948 turf battle over aircraft that was resolved by a document called the Key West Agreement, the army was banned from owning any helicopter weighing more than 4,000 pounds.* Even though the air force gave way on this ban later, the army was the last service to receive S-55 transports for the war.

The marines were the first to experiment with helicopters for combat operations, and encouraging signs appeared by August 1950. The first marine helicopters belonged to VMO-6, an observation squadron that ended up doing much more than target spotting. They made the news so frequently that readers must have thought the unit owned dozens of helicopters; in fact the squadron had only four. Their Sikorsky R-5 Dragonflies flew constantly, making rescues in the Pusan perimeter.

The navy put snipers on their helicopters and tried clearing mined harbors with them. Other snipers took aim at infiltrators sneaking across the front lines. The marines dropped in special-operations teams via helicopter. One air force helicopter raid carried out the remnants of a Soviet MiG-15 fighter.

The helicopter's newfound popularity was already having a dizzying effect on the tiny industry. Helicopter makers had gone from depression to euphoria over their rapidly filling order books, and were now verging on desperation about how to meet those orders.**

How would the new war horses hold up in combat? No one knew: some parts of the helicopter were highly vulnerable but maybe enemy gunners would miss them entirely. According to Burnet Hershey, author of the 1943 book *The Air Future,* "The helicopter's easy maneuverability gives the pilot almost complete control of the air, and with that control, the ability to dodge gunfire." But

* This suited the air force, which wanted to control tactical air support and transport, thus limiting the army helicopters to evacuation and battlefield observation. Accordingly, the air force asked manufacturers to create very large helicopters, comparable to the C-47 fixed-wing transport, which it planned to use in moving supplies and men to the front. The giant Piasecki H-16 Transporter, later discontinued after a crash, was one product of that directive.

** One example is Kaman Aircraft. At the start of the Korean War, the helicopter maker had twenty-five employees on staff. After less than two years, the employee roster was up 4,500 percent. This even though no Kaman helicopters were ever used in the war.

dodging was not possible for helicopters coming in to make battle-field rescues when enemy gunners ringed the site. The first helicopter crash in combat came in September 1950, when Marine Lieutenant Arthur Bancroft was shot down while attempting to rescue a fixed-wing pilot north of Seoul with his Sikorsky R-5 Dragonfly.

Rescues were important for troop morale and commanders reveled in the battlefield view they provided, but to marine tacticians of the era the greatest contribution of the helicopter was to transport large numbers of troops to storm an objective. While parachutes and transport planes certainly could deliver a huge number of troops behind enemy lines, the helicopter's advantage lay in getting smaller numbers to specific destinations and then retrieving them.

The Sikorsky S-55 (also known as the H-19, HRS-1, H045, and Chickasaw) was the principal American troop carrier to go into battle in Korea. It had originated with a bid to win an air force competition for an Arctic rescue helicopter. While Piasecki won that contest, the air force saw other possibilities for what became the S-55 and began ordering them anyway, in small numbers. In profile the S-55 had a faintly whimsical appearance, with its fuselage shaped like a loaf of bread on wheels. Pilots sat high up in the machine, over the front-mounted engine, offering a large space for passengers and cargo near the center of the aircraft.

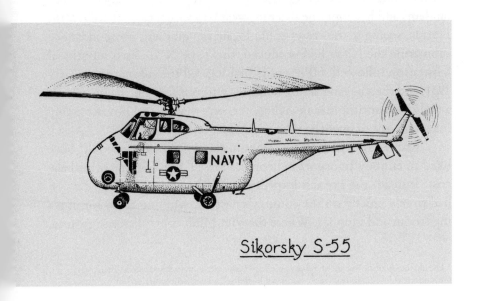

Sikorsky S-55

This aircraft, forty-two feet long and with room for up to ten passengers, introduced a new way of fighting. By mid-September 1951 helicopters had proven they could bring ammunition, food, and reinforcements to marines in remote areas, but could they move entire units around the battlefield? If so, it would be the biggest improvement in mobility since massed parachute drops.

A full-scale trial came on September 21. Commanders picked out a 3,000-foot, steep forested ridge on the eastern coast, a four-day hike from friendly lines by foot, and assigned twelve Sikorsky S-55 helicopters to the job. Three helicopters hovered to let down an advance team on ropes, which knocked down trees for a landing zone.* Just four hours was enough time to deliver 228 men and nine tons of supplies to an outcrop dubbed Mount Helicopter (Hill 884 on the maps). North Koreans had the area surrounded but were unable to shoot down any of the aircraft. The tyranny of terrain, it seemed, was over.

Within a year such successes made it clear that the Key West Agreement was pointlessly blocking the army from getting the big transports it needed to move men and supplies. The air force agreed to divert a share of Sikorsky's output to the army. In 1953, after three years of war, the army received its first S-55s. Even the British had gotten them earlier.

The navy wanted specialized models, one priority being submarine hunting. Beginning in 1950, the navy had put Piasecki HRP tandems and hydrophone listening gear to a severe test off Key West, Florida, seeing if the crews could locate the quietest, most advanced submarine the Nazis had produced, the Type XXI. Two months of frustration followed. HRP crews first located the submarine in early 1951, and then began finding it almost without fail. The navy ordered a dedicated antisubmarine helicopter, one that would hover reliably for long periods over the ocean's surface.

France, then fighting Communists in Indochina, would have liked to employ transport helicopters as well but couldn't afford the cost. Instead, the French Expeditionary Corps and Foreign Legion had to rely chiefly on the country's narrow, winding roads for moving troops and supplies. Where the Viet Minh controlled the ground,

* During the Vietnam War such a team was called a Pathfinder unit. Pathfinders cut down trees, probed for enemy combat teams around the perimeter, and guided helicopters' arrival and departure.

the cost of highway travel could be very high, such as the calamitous ending to Operation Lorraine. In October 1952, the French sent 30,000 troops, supported by tanks, deep into enemy territory from the French-fortified positions along the de Lattre Line. Their job was to attack Viet Minh supply storage dumps 100 miles away in the Phu Doan area. The French expected this incursion would force the Viet Minh to stop attacks elsewhere and perhaps bring their massed army out into the open, where French airplanes could attack them. After a few weeks the massive ground raid had reached Phu Doan and located hundreds of tons of arms, including much new Soviet equipment. It had not, however, brought Viet Minh divisions out into the open, where the French could use World War II tactics. Having taken the towns, General Raoul Salan decided not to hold them any longer; he was now worried that Viet Minh commander General Vo Nguyen Giap might have laid plans to attack his overextended column, which was 100 miles from the nearest French lines. A trap laid for columns in narrow defiles where the enemy could not bring its power to bear was one of Giap's trademark moves. In mid-November Salan ordered the men to head for safety but, as he had feared, Giap's men were waiting.

The Viet Minh anticipated that the French in pulling back from Phu Doan must pass through the steep-walled valley at Chan-Muong. Here they built log roadblocks and set up heavy weapons trained on its narrowest point, a bottleneck formed by steep rock walls and manioc fields on either side of the road. On November 17 Viet Minh battalions let a line of tanks pass through mostly unscathed, then attacked the more vulnerable middle of the column. Exploding vehicles in the rear blocked help from tanks that had yet to enter the defile, and exploding vehicles in the front blocked escape in that direction. Mortars and hand-placed satchel charges destroyed each trapped vehicle. French commanders gathered up their infantry from the confusion and sent them up the slopes to attack the mortar and gun positions. A bayonet charge finally sent the Viet Minh into retreat. The battles continued for another week, until the column reached safety at the de Lattre Line. The cost was 1,200 casualties. If the French had had helicopters, it seemed to some observers, such ambushes might be a thing of the past.

But France was on a tight budget and never fielded more than a handful of helicopters.[1] These were reserved for battlefield observation and evacuation of the wounded. Evacuation flights began in May 1950 using two Hiller 360 light helicopters. In the kind of improbable story that wars plow up, a brain surgeon came to personify the benefits these machines could bring to wartime medical care.

Her name was Dr. Valérie André. After completing her training as a neurosurgeon in 1948, she volunteered to serve with the medical team supporting France's paratrooper forces in Indochina; this often required her to parachute into combat zones to work on cases of brain injury. She found parachute drops such a frustrating form of medical care that she went back to France in June 1950 and learned to fly helicopters. "Thanks to them we now would be able to bring the wounded back to the rear, to heal them quietly and safely, instead of having to drop our medical team close to the theater of operations," she recalled in a speech later to the Whirly-Girls association of women helicopter pilots.

After twenty hours of flight instruction, André returned to Vietnam late in 1950 and began a new career as a combination of medevac pilot and brain surgeon, usually operating out of an air base near Hanoi. By the time André returned to France in 1953, she had flown 165 soldiers from the battlefield.

André also flew medical evacuation helicopters in the French war against Algerian rebels that began in 1954. By 1956, the French were operating ninety helicopters there and placed orders for more. In the end they employed over 600. Unlike the very limited role for helicopters that the French had assigned for helicopters in Indochina, here the French used them in full range of conventional warfare and special operations, including the relief of besieged troops at Bab Taza Pass in 1955.[2]

French infantry commanders rejoiced over the way helicopter mobility broke up the usual maddening pattern of rebel sneak attacks in mountainous areas. Before helicopters, rebels would pick an ideal ambush site, shoot up the convoy, then pick off the French soldiers who struggled up the rocks to get at them. After inflicting many casualties, the rebels would melt into the rocky reaches. Such fighting was simple and very effective.

The French responded by posting troop helicopters near known ambush points, so that when such a battle began they could place a cordon of troops on ground even higher than the rebels occupied. Then legionnaires closed the ring. "Vertical envelopment" recalled the American Civil War, when commanders sent off horse troops behind enemy lines to cut supply lines and attack command posts.

The British had their own antirebel action under way on the Malayan Peninsula. This conflict also suggested that helicopters might improve war-making radically. The British military success in this 50,000-square-mile country depended in part on the mobility provided by Sikorsky S-55 and Westland Whirlwind helicopter transports operated by the Royal Air Force and Navy. In that conflict, Communist insurgents, mostly of Chinese origin (called CTs by the British, short for Communist terrorists), had tried to drive British economic interests off that rubber- and tin-producing peninsula with a campaign of terrorism and later insurgent warfare. Even with their comparatively small force, the British forced the CTs to give up the fight by 1960. Helicopter-borne troops surrounded the enemy forces, protected outposts during attacks, and evacuated 5,000 wounded troops plus hundreds of sick citizens as well. In the words of Sir Gerald Templer, the Tiger of Malaya, the British had won not just the battles but also "the hearts and minds" of the people.

So helicopters had performed well in wars of the 1950s, and planners could reasonably think helicopters would do even better on behalf of technological nations in the wars to come, because the aircraft were getting bigger and were carrying more precise and powerful weapons. The U.S. Army Air Corps considered a proposal to mount bomb racks on helicopters during World War II, but did not pursue it. Interest in armed helicopters revived in the 1950s: while fixed-wing fighter-bombers could carry heavy loads of air-to-ground munitions and could reach the battlefield quickly, their speed made them inaccurate and their fire was intermittent. Helicopters would be better at picking a target and delivering fire.

Experiments at Bell Helicopter began in 1951, when Joe Mashman and Hans Weichsel of Bell visited the Pentagon to borrow a bazooka. Bell wanted to test the Model 47 for its ability to harass the enemy on the ground, and also wanted to make sure that the anti-tank weapon's rocket exhaust wouldn't damage the machine.

It was a simpler time. Stewardesses allowed Mashman and Weichsel to bring their bazooka onboard, as carry-on baggage, on the airliner flight to Buffalo. They tested it at a range in Virginia. Despite the flame and smoke that followed each firing, the helicopter sustained no damage, as had one helicopter in December 1945 when tested with a recoilless rifle.

The French were the first to take armed helicopters seriously, after finding that losses of troop-carrying helicopters to ground fire were unacceptable unless other aircraft were on hand to fight off Algerian rebels in the rocks. Colonel Felix Brunet first mounted machine guns and rockets on Sikorsky S-55 transports, and went on to arm other models.

Armed helicopters proved much more effective than skeptics had predicted. Even after the Algerian lessons started coming back, some American military experts continued to predict that the best the helicopter could do for commanders was to reprise its role in Korea, as troop carrier. As a *New York Times* reporter summarized the conventional thinking in 1960, "The value of suppression fire by armed helicopters against ground areas is probably of marginal importance."

The first guided missile to go into action from a helicopter was the French AS-11. It was part of an overall strategy of creating an interdiction zone, or no-man's-land, across the eastern and western borders to cut Algerian insurgents off from their supplies. With two infantrymen lying in the ambulance-litter baskets and firing forward, it also served as a platform for automatic weapons. Thus armed with guns and antitank rockets, a new helicopter from France called the Alouette II began shooting up Algerian rebels in caves. The Alouette II also received notice worldwide because it was the first helicopter used in war powered by a gas turbine instead of a piston engine.

The shortcomings of gasoline engines inside helicopters had long been the subject of unhappy comment by pilots and commanders. Following a military-aviation report of 1954 that analyzed the usefulness of helicopters for transport, Pentagon representatives summoned the manufacturers and told them that helicopters needed more payload capacity, and had to be more rugged and easier to repair. One point stood out: lose the piston engine. Even the best gasoline-fueled engines were heavy and bulky for the power they offered, taking up much of what would otherwise be payload capacity, and required much field maintenance.

Shortly after World War II ended, two approaches appeared to hold promise as alternatives to the piston engine. One was the tipjet and the other was the gas turbine.

The tipjet shot pressurized gas backward from the tips of the rotor blades, similar to the "aelopile" toy of Greek mechanical inventor Heron of Alexandria. The aelopile drew steam from a kettle of boiling water, and sent the steam through a hollow axle into a hollow ball, where it left by nozzles. The escaping steam, aimed opposite to the direction of rotation, caused the ball to spin on its axle. Inventor W. H. Phillips had used the tipjet principle for his model helicopter more than a century earlier. If the rotor blades are left hollow like a pipe and have nozzles on the end pointing at a right angle, either compressed gas or fuel can flow from the fuselage down the blades. As gas or flame shoot out the nozzle, the blades will turn and generate lift. Austrian inventor Friedrich Von Doblhoff built the first full-sized helicopter of this type during World War II; a pump on Von Doblhoff's machine compressed air and sent it along the blades to emerge at a nozzle.*

Hiller knew the tipjet was a risky detour from the proven gasoline engine, but it offered great simplicity. It could be the price breakthrough that the family helicopter-runabout needed to get off the ground and into mass production.

Unlike Von Doblhoff's approach, which emitted no flame from the rotor blades, Hiller's plan was to put a combustion chamber on the blade tip. That way it could offer more thrust. In 1947 Hiller began testing tiny jets on the ends of rotor blades, first of the pulse-type jet that powered the V-1 missiles that bombarded London, and then of the ramjet type.**

In Hiller's model, a one-horsepower, lawn-mower-type gasoline

---

* Sud Aviation of France used a similar approach for its Djinn helicopter. The Djinn was the only tipjet helicopter to reach full production and sale. Sud discontinued it because of low flying speeds and high fuel consumption.

** Ramjets were the 1913 invention of Rene Lorin, who realized that if air flowed quickly enough into a combustion chamber and was compressed in passing through, and if fuel and flame were added at that point, it would amount to a rocket with no moving parts.

engine accelerated the rotor blades to working speed, after which the two ramjets (weighing only twelve pounds each) took over. The ramjets needed only ignition and kerosene fed from a tank in the body of the helicopter. Such a helicopter didn't even require a tail rotor, since there was no engine mounted on the body of the helicopter to generate unwanted torque. When flown for the press in 1951, the Hiller Hornet looked like a cartoon-copter come to life, quite small, resembling a streamlined telephone booth with a short tail boom attached and resting on tiny wheels. Hiller was determined that the aerial flivver should fit in the average garage, so he arranged that the tail boom could come off with removal of a few wing nuts. It had seats for two people. Hiller predicted a retail price of $5,000 and promised a range of fifty miles. Sportsmen could use it.

Early in the Korean War the military needed huge numbers of conventional helicopters rather than a handful of experimental ones, but the army and navy agreed to buy Hornets in 1952 for testing as an artillery observation aircraft transportable by truck. If it worked, one or two soldiers could check out enemy positions and direct fire by

Hiller Hornet

walkie-talkie. By 1954 Hiller had a ramjet-driven helicopter that he
felt was comparable to a miniature Jeep and suitable for use by both
military and civilian pilots. The army brought two of them to Fort
Rucker, Alabama, to test for usability as a gunship. The navy ran its
own tests in Maryland.

The army's Hornet was truly memorable at night. Though tiny,
it made a thirty-foot-wide ring of fire in the air and roared "like a
Boeing 707," recalls Link Luckett, who saw it in action during tests at
Fort Rucker. Residents living near the original ramjet test site at the
Hiller factory in east Palo Alto had been making the same noise com-
plaints for years.

The heavier military model emptied a full tank of fifty-five gallons in a half hour, meaning the craft barely had time to leave town
before it had to start sniffing for fuel.[3] Pilots reported that they could
make safe emergency landings if the two engines quit, but only if
they had learned to autorotate the Hornet way, by starting the landing flare much higher than usual. Not surprisingly, the army and the
navy lost interest after buying less than two dozen test articles. One is
visible still at the Smithsonian's Steven F. Udvar-Hazy Center, in the
far corner reserved for helicopters. One selling point for the Hornet
was that it did not need the number of high-maintenance parts that
other helicopters required. The tipjet engine could be changed out
with just a screwdriver. And to be fair, the early model civilian
Hornet performed better than the military models, being lighter and
smaller.

After he realized that his first ramjets were proving too ineffi-
cient and too low in power, Hiller continued to experiment with
other tipjet designs, one using highly unstable "high-test peroxide"
as a propellant. In this system, hydrogen peroxide of at least 87 percent purity is sent through a fine screen of silver metal, which instantly turns the liquid to oxygen and hot steam. Gilbert Magill's
one-man Pinwheel helicopter also relied on this liquid. Like the
Hornet, the Pinwheel was thirsty for fuel, burning sixty gallons per
hour, but was able to lift a single man into the air. While posing for
photos taken in 1957, Pinwheel pilot Richard Whitehead wore a
strained expression, perhaps because the rig weighed 150 pounds and
his legs served as landing gear.

The tipjet had one other advantage: It could lift great weights as long as no great distances were involved. This raised the possibility of constructing mammoth helicopters able to lift hundreds of tons, big enough to snag Apollo spacecraft boosters for NASA as they parachuted down.[4] Hughes Aircraft built a tipjet-powered helicopter called the XH-17 and invited the press to see a nine-minute demonstration flight of what one reporter called a "noisy, flame-spitting monster that did everything but fly upside down." It was historically interesting not for the long-term prospects of that helicopter but for the fact that the October 23, 1952, demonstration was the last time tycoon Howard Hughes ever appeared before a group of reporters.

So the tipjet worked, and made a spectacular show, but its thirst for fuel was never solved. Fortunately for aircraft designers, a much more productive line of engine development was ready.

It was the gas turbine, an old idea with a gestation period almost as protracted as that of the helicopter itself. John Barber of England designed a turbine in 1791 that would have used exhaust gases from burning coal to turn wheels and provide power. No working gas turbine followed, however, until the *Société Anonyme des Turbomoteurs* got one working in 1906. Over the ensuing decades, gas turbines remained too big and heavy for doing anything but running pumps or electric generators. The demands of war changed that, and a new generation of light and powerful turbines swept into the helicopter business in 1950.* One prompt beneficiary of these new powerplants would be the UH-1 Iroquois. The model began test flights in 1956 as the Bell XH-40, and was intended to be primarily a battlefield-evacuation aircraft. Its engine was the Avco Lycoming T53, the lineage of which traces back to a team of mechanical engineers who produced the engine for Germany's jet fighter, the Me 262.

A turbine for a helicopter takes air in the front, compresses it greatly with a fan, adds a mist of fuel, moves the mixture back to a combustion chamber, ignites it, and shoots it past a set of bladed wheels that harvest much of the gases' expanding power. Gas turbines need no radiators or conventional cooling system. The bladed wheels, attached to reduction gears, drive the rotor system.[5] When perfected, helicopter gas turbines generated as much power as a pis-

---

* The first turbine helicopter to take flight was the French-built Ariel III in April 1950, using a Turbomeca Artouste turbine engine. That demonstration was followed the next year by the Kaman K-225 prototype.

ton engine weighing four times as much, and needed only half the maintenance. Altogether they gave helicopters radically more range and payload.

So, rising up to look over the landscape in 1959, one year prior to the Vietnam War: the helicopter had been in regular use for two decades. Approximately 8,000 had been built in the United States. It had been but a supporting actor during military conflicts and for urban transportation, but turbine power looked to change that within a few years. At Fort Rucker in Alabama, General Carl Hutton was building a case for a new concept in warfare called "sky cavalry."* Skilled pilots were graduating from army aviation programs at that post and at Camp Wolters, Texas. Still, with no war in sight, helicopters looked likely to make a bigger mark in the civilian world than the military. Shuttles operated large helicopters in three large cities, sustained by federal mail-hauling subsidies.** Proposals were circulating for downtown heliports that would receive dozens of helibuses per day. A few newcomers to the personal-helicopter business still held out hope of selling thousands of such craft, but there was no indication that costs would ever drop low enough to lure the average suburbanite. Would Americans ever see a place and time in which helicopters flew with the ubiquity of station wagons? Indeed they would, but not in America.

* The Sky Cavalry concept, as delineated during the massive war game called Operation Sagebrush in 1950, was to retain fighting forces at dispersed points, then use helicopters to apply that concentrated power at critical points at the critical time. This would be followed by rapid dispersal. The tactics were driven by commanders' belief that a future enemy would use tactical nuclear weapons; therefore, infantry forces should never be concentrated except when necessary.

** The total of city-based, scheduled helicopter airlines in the U.S. since 1947 is less than twenty. The busiest ones operated in Chicago, New York, Los Angeles, and the San Francisco-Oakland area.

Chapter Ten
Operation Chopper

In early 1961 General Maxwell Taylor, special military adviser to the White House, recommended to newly inaugurated President John F. Kennedy that the United States begin offering air transport and in-the-field military expertise to the Army of the Republic of Vietnam, or ARVN. The U.S. was already lending equipment and advisers to fight Communists in Laos, therefore doing the same inside Vietnam did not appear to pose a slippery slope toward war. Despite some misgivings from the Pentagon, President Kennedy endorsed the idea. Accordingly, in December the escort carrier USNS *Core* tied up in front of the Majestic Hotel on the Saigon riverfront; aboard were nearly three dozen of the army's piston-engine, sway-backed H-21C transport helicopters, popularly known as Flying Bananas. It was the first major military support provided by the United States for Vietnam.* They went into action one week before the

---

* But not the first use of American helicopters in the new war. Three months earlier an American-provided Sikorsky S-58 Choctaw had flown into Laos to rescue a wounded ARVN commando after a cross-border raid.

end of the year. In Operation Chopper, thirty-two American pilots flew a thousand ARVN soldiers into battle. The troops surrounded and captured an enemy radio transmitter.

That code name could serve as a name for the whole war. "Our tactics in Vietnam were based on massive use of helicopters," General William Westmoreland, commander of the United States forces in Vietnam, wrote. "What would we do without helicopters? We would be fighting a different war, for a smaller area, at a greater cost, with less effectiveness. We might as well have asked: 'What would General Patton have done without his tanks?' "

For every helicopter the French used in Indochina, the United States used at least three hundred. At first these were piston-powered ones, greatly underpowered for the conditions. In addition to the H-21 Bananas carried by the *Core,* these included the widely used, loaf-shaped Sikorsky S-58 Choctaw* transport and the nimble Bell OH-13 Sioux and Hiller H-23 Raven light helicopters. Each succeeding year the scope of helicopter work grew. They carried people and mail between bases on "ash and trash runs," which cut risks from ambushes and roadside bombs. Helicopters dropped in troops, picked them up, and served as a platform from which commanders could observe the action and give directions. Big helicopters raised little ones from the jungle to bring them back for repair. Marine sharpshooters rode helicopters to strike down the enemy by night. Helicopters dispensed defoliant and propaganda leaflets. Helicopter traffic rose to a volume that has never been equaled and likely never will be.

Broadly speaking, helicopter use in Vietnam passed through three principal roles, two of which are covered in this chapter. In order of appearance, those roles were troop transport, air assault, and large-scale evacuation.

From the first, helicopters provided rapid transit for troops, who typically were attempting to employ a "cordon and sweep" maneuver to round up the enemy. Soon, using tactics borrowed from Algerian rebels, the North Vietnamese learned how to shoot helicopters down. Americans tried new tactics and new helicopters. While most air-assault missions failed to find and fix the

---

* S-58 is the most common name for this model. The army's version was the H-34.

enemy, there were some units who registered considerable success before the United States began pulling out of Vietnam in early 1969. Their successes came in part from using the new helicopters in new ways, ways that, at the same time, were grounded firmly in lessons from pre-helicopter battles. One such commander was Hank "Gunfighter" Emerson, of Milford, Pennsylvania,* who as an army infantry officer organized offensives in the Central Highlands and later the Mekong Delta. His innovations were not heavily publicized at the time, nor featured in popular treatments of the war afterward.

I first came across references to the Gunfighter story while reading transcripts of songs taped during talent shows by army aviators during the war.[1] Many were musical complaints about foolhardy missions, clueless commanders, mechanical malfunctions, and mistakes committed by novice pilots. Some songs took aim at the growing fame of the Green Berets, dubbed Sneaky Petes.

Since commanders usually come off badly in any song or skit drafted by combat troops, one song stood out. Titled "Gunslinger," it referred to a lieutenant colonel named Hank with the 101st Airborne Division. In the song, the colonel's helicopter is shot down after enemy fire cuts the hydraulic lines. Rather than hunker down for rescue, Hank gathers his men, strikes up a song, and goes out to chase down the Victor Charlie who shot at him. The enemy replies with hand grenades, but the colonel roars his defiance and is victorious. He opens up two canteens of whiskey and shares them with the troops as they await a lift out.

"I never had a better time or met a better gang," the song concludes, "The night we whooped old Charlie's ass then partied in Phan Rang."

"The unusual part of all of this is that 'Hank' was not a pilot or crew member," remarks Marty Heuer, a former officer of the 174th Assault Helicopter Company who helped gather the song archive. According to Benjamin Harrison, then a brigade commander in Vietnam with the 101st Airborne, the songwriter clearly was refer-

---

* Emerson, the son of an army surgeon, was born at Walter Reed Army Hospital. His family tree had four successive generations of doctors and he was to be the fifth.

ring to Colonel Henry Everett Emerson, known as "Gunfighter" to his men. I located Emerson in Montana.

Emerson's first brush with the long war in Indochina came in March 1953, as a military aide to General Mark Clark. Dwight Eisenhower had asked Clark to bring back a personal report. Emerson, who by then had experience as a platoon leader and company commander for the Fifth Regimental Combat Team in Korea, spent hours swapping stories with the officers of the French Expeditionary Corps. The French described how the Viet Minh were cutting up their truck convoys with ambuscades. "They had the flower of the French Army there," Emerson says now. "These were Legionnaires, and first rate, but they were having a hell of a time.... Clark sent a letter back, saying, 'This is a morass—we want no part of this.' And Ike stayed out of it. The terrain was all wrong, and there were long logistics lines."

The man in charge of giving the French that hell was the remarkable General Vo Nguyen Giap, entirely self-educated in tactics, born to an educated peasant in northern Vietnam in 1911. In 1941, while hiding in a cave in the limestone mountains of Pac Bo on the Chinese border, Giap had helped ignite the Viet Minh liberation movement. He organized an attack on a French military outpost in December 1944 that the Vietnamese came to regard as the opening shot by their revolutionary army forces. It generated a huge boost in confidence. Still, the Viet Minh war against the French saw no great progress until the Chinese began sending large quantities of arms and ammunition in 1949.

When Viet Minh forces besieged and overwhelmed Dien Bien Phu in 1954, Giap was in command of those Communists. When the U.S. helicopters arrived on the *Core* in December 1961, Giap was in command. When the U.S. pulled out in 1973, Giap was still in command.

March 1962 saw helicopter "sweeps" along the Cambodian frontier in An Giang Province. More army assault helicopters arrived in 1962, along with the first marine helicopter squadron. Under pressure from Defense Secretary Robert McNamara to use helicopters more than already planned, the army arranged for tests of helicopters on a much larger scale, which would free troops from primary dependence on truck transport. In time this would lead to air assault divisions and air cavalry combat brigades.

Casualties were not long in coming. The Vietcong shot down their first U.S. helicopter shortly afterward, west of Saigon. This crash and other single losses could be written off as lucky shots, but not the incident at Ap Bac, a year later. The U.S.-advised plan on January 2, 1963, was to have ARVN troops of the Seventh Division attack a Vietcong command post near the village of Ap Tan Thoi, southwest of Saigon. American pilots in ten Flying Bananas would fly in ARVN troops. More ARVN would arrive on armored personnel carriers, by truck, and by riverboat. Five armed Hueys orbiting overhead with rockets and machine guns would cover them.

But many more Vietcong than expected were waiting in and around Ap Tan Thoi. They had heard of the attack through spies among the ARVN and in American military bases, and using a new tactic had dug themselves in at the nearby village of Ap Bac, along with Soviet DShK 12.7-millimeter heavy machine guns, which though old-fashioned were extremely dangerous to helicopters at close range.

After the first waves of ARVN troops came under fire, American advisers suggested that additional troops be flown in from uncommitted reserves. Tragically, they picked a landing spot that happened to be near a Vietcong strongpoint; therefore these reinforcements were exposed to the heaviest concentration of mortars and machine-gun fire. Shrugging off the Allied artillery, the Vietcong shot down one Flying Banana immediately, then a second one that came to rescue the stranded crew. Two more helicopters got away, but with damage. As Hueys fired into the tree line, the crew of one Huey attempted a rescue of the first two American crews and was shot down. When armored personnel carriers moved forward, the Vietcong sent a squad into the open to attack with Soviet-designed rocket-propelled grenades.*

By the time the Vietcong withdrew, two more helicopters had crashed at Ap Bac, for a total of five. Of the fifteen helicopters that

---

* The Vietcong also found the RPG-2 weapon and its successor the RPG-7 to be effective against helicopters. Over the course of the war these cheap, unguided warheads accounted for 10 percent of all American helicopters brought down by ground fire.

participated, one returned without damage. The North Vietnamese press trumpeted the news as evidence that the Vietcong could meet the enemy head-on and win. A new set of tactics circulated among the Vietcong commanders, based on lessons from Ap Bac: attack remote posts as a provocation, allow them to radio for help, and position heavy machine guns to meet the inevitable counterattack by helicopter. "It's better to stand and die than to run and be slaughtered," was the well-publicized conclusion of one Vietcong infantry commander.

Top U.S. officials told the press that Ap Bac was just another in a long string of "win some, lose some" events, so there was no special meaning to the battle. Military adviser John Paul Vann was shocked by the losses and by the performance of the ARVN troops, believing that the ARVN had failed to counter the attack as directed and to cordon the Vietcong.[2]

Ap Bac also generated a distinctly more skeptical tone in news coverage. Reporters, including Neil Sheehan, David Halberstam, and Peter Arnett, took the losses at Ap Bac as confirmation of their suspicions of the previous year, that the U. S. military had been covering up both its own shortcomings and the fighting ability of the Saigon regime. In August 1963, Halberstam wrote that the United States must be losing ground in the Mekong Delta, because Vietcong battalion strength and armament stocks were growing there.

The havoc wrought upon American helicopters at Ap Bac was no surprise to historian Bernard Fall, author of a classic text on the French-Indochina War titled *Street Without Joy*. Fall knew that a North Vietnamese military delegation had visited Algeria in April 1961 to learn how to fight Western mechanized forces. Months later the Algerians came to North Vietnam to continue the exchange of information; and in December 1961 the North Vietnamese began broadcasting lessons over the radio on how to identify and destroy American helicopters.

There would be more battles, with heavier helicopter losses, particularly after Congress authorized direct military action in August 1964.[3] American combat troops and helicopters began making major offensives in the summer of 1965. Operation Samurai IV in the A Shau Valley in 1965 saw seven helicopters lost in one day.

Vietcong forces began targeting the machines at rest: at Chu Lai, sappers blew up nineteen helicopters with satchel charges in a single night in October 1965.*

But a month later helicopters helped revolutionize a battle at Ia Drang, to the Americans' advantage. This battle arose out of the fact that the North Vietnamese Army was making a major push across South Vietnam. At Ia Drang, the First Cavalry used assault and transport helicopters in defeating three regiments of the North Vietnamese Army: shifting entire battalions around, moving artillery from hilltop to hilltop, and hauling in 13,000 tons of supplies during the weeklong battle. The bad news went back to Hanoi: Because of helicopters, the Americans could appear almost anywhere, and in force. But commanders' reports added that most of U.S. units appeared to be so dependent on helicopter travel and resupply that they lacked close knowledge of the terrain and the people. When villagers provided assistance to the NVA and Vietcong, American troops had difficulty knowing or doing anything about it.

The stiff resistance that the Vietcong and NVA put up against technological warfare suddenly boosted prospects for the original helicopter medevac unit in Vietnam, the Fifty-seventh Medical Detachment. The Fifty-seventh had begun carrying out wounded in May 1962 but faced opposition in the next two years from infantry officers who felt that the unit's five unarmed helicopters could be better used in combat, and should be tapped as a source of spare parts if not commandeered entirely. Its first base near a field hospital at Nha Trang Air Base in the north was too far from fighting, which was then concentrated in the south. But the battle at Ap Bac made the point that medevac helicopters must be staged close to the battlefield. The unit carried out almost 2,000 evacuations by the end of 1963.

Major Charles Kelly took over command of the Fifty-seventh Medical Detachment in January 1964, which was also the same period in which the unit acquired the permanent radio call sign Dustoff. The fearless Kelly expanded the unit's reach in both geography and time by authorizing recoveries after dark. According to

---

* By one estimate, a third of the army's helicopter fleet in Vietnam was damaged or destroyed each year after such warfare began in earnest. Noteworthy was Lam Son 719, when ARVN forces crossed into Laos. American helicopters supported that incursion. Using Soviet SA-7 missiles and radar-guided antiaircraft cannons, the NVA destroyed 150 helicopters in two months.

Dustoff lore, Kelly told his men that they should not refuse any evacuation request. Responding to a report of wounded men in an ARVN unit on July 1, 1964, Kelly declined to turn back despite radio messages from the American adviser on the ground telling him to abort the mission because the volume of fire was too heavy. "When I have your wounded," Kelly told him. Kelly was fatally shot while searching for the casualties. Jay McGowan served as a Dustoff pilot in 1964 and 1965, flying out of Tan Son Nhut Air Base for the Eighty-second Medical Detachment. "We didn't fly into these situations blindly," he says. "There's no sense having four men killed to get one man out."

"I admired the medevac helicopters a good deal," says retired anthropologist Gerald Hickey. Hickey spent seventeen years in Vietnam beginning in 1956, studying the ethnography of South Vietnam from the remotest highlands to the fertile rice-growing areas of the Delta. "Those helicopters were a godsend during the war." He says that while the army-trained pilots looked to be as young as teenagers, they knew their aircraft and were absolutely fearless. The youngsters drove their choppers like hot rods, landing in the middle of firefights when pilots from other services might defer. Hickey was visiting[4] Camp Nam Dong in July 1964 when enemy soldiers attacked overnight. The defenders* held the perimeter against an estimated 900 attackers through the night, but enemy forces remained in the vicinity the next day, and their machine guns drove off the first attempt of marine S-58 helicopters to deliver food and reinforcements. Shortly afterward a single Huey, door guns blazing, cleared a path for the relief helicopters. Hickey examined the enemy dead and advised U.S. commanders that they looked like visitors from the North rather than area residents. Hickey's insight into their identity was one of the first indicators that the North Vietnamese were beginning to invade.

Clearly, U.S. forces at Ia Drang proved they could bring devastating firepower to bear at more battles like Ia Drang, but such opportunities proved to be rare. What good was a helimobile force that struck like so many lightning bolts if the enemy refused to come out

---

* Including Gerald Hickey, Ph.D., who wielded an AR-15 and was wounded by an explosion during the fight.

and fight fair? Were American forces incapable of driving the enemy out of South Vietnam?

Such questions were timely to Lieutenant Colonel Hank Emerson. In 1965 Emerson was at the U.S. Army War College at Carlisle Barracks, Pennsylvania, preparing for his first tour in Vietnam as a battalion commander for the 101st Airborne Division. Feeling that the standard curriculum would not suffice, on nights and weekends he burrowed into the library. Emerson read books by Mao Zedong and Vo Nguyen Giap. He read about the French-Indian War and border wars on the American frontier. "I was most interested in guerilla tactics," Emerson recalls now, "where the enemy fights like Indians." Early in his military career Giap had similarly spent many days reading classic military literature, working out tactics against French and then Japanese occupiers.

One of the books of special interest to Emerson was *Shoot to Kill*, by British infantry commander Richard Miers. Miers explained how, beginning in 1951, the British worked out tactics to surround insurgents based in the Malayan jungles. Four platoons radiated from a common center. As the tactic evolved, the British Ferret Force began using helicopters to surround and capture insurgents. One key helicopter unit was the Royal Navy's 848 Squadron, whose feisty motto was *acip hoc*, Latin for "Take that!" The British had progressed beyond the truck-minded use of helicopters to employ them in setting up a cordon, or ring, with which to trap insurgent forces in the jungles of Pahang for capture or aerial bombing. Later this would be called vertical envelopment. Guided by the commander in an orbiting helicopter, Westland and Sikorsky helicopters of the 848 Squadron shifted infantry units to surround the enemy, using jungle clearings when possible. When the troops needed to make a clearing, they slid down on ropes and cut the trees with saws and the head-high grass with parang knives. By 1952 the British under Gerald Templer had insurgent leader Chin Peng on the run, having killed four of his top commanders.

Intrigued by the lessons that the British had worked out during their years in the Malay jungles and rubber plantations, Emerson wrote a term paper proposing a set of infantry tactics for Vietnam that grew out of lessons from the Ferret Force, but also proposed some changes.* For one thing, he says now, the Ferret Force's "weakness was being tied to a base, and the enemy would learn that base pretty soon." After leaving the War College, Emerson tried out his tactics during field exercises at Fort Pickett and in the Dominican Republic.

As Emerson took command of his battalion[5] in October 1965, the forested north-central region in which he would operate (labeled II Corps and III Corps on American maps) was serving as a slow-motion highway for North Vietnamese invaders who passed through on foot. It was a year of rapidly escalating conflict. Seemingly the entire NVA was migrating toward the rice paddies and cities of the South, ten to fifteen men at a time. He had orders to find and destroy them.

Emerson's paper laid out an approach he called the Checkerboard. While its use varied greatly according to terrain, it relied on three basic elements. The first was infantry: three to four rifle companies from his battalion, broken into smaller units. All units kept in touch by radio. There were Recondo squads who served as scouts, and smaller groups that maintained observation posts by which to monitor enemy movement. Typically Emerson's battalion kept one rifle company in reserve.

The second element was a very restricted use of helicopters to move infantrymen around the area of operations. Whenever possible, "we all walked," says Thomas H. Taylor,[6] one of Emerson's company commanders. "We didn't want to insert by helicopter because the NVA would know our location. We'd slip through the jungle and meet at a trail junction or near a rice cache." The third element was artillery positioned on hills or in clearings that could lay down a steady stream of shellfire for miles in all directions.** Helicopters shifted the howitzers as necessary.

In Emerson's original conception, these elements were to play out across a checkerboard on which U.S. troops would move like

* Emerson's term paper, which is still cited in manuals of the 101st Airborne Division, was titled "Can We Out-Guerilla the Communist Guerillas?"

** Infantry commanders valued the support they received from well-aimed howitzers, one reason being that artillery was unaffected by cloud cover.

game pieces, choosing their own moves. While in practice Checkerboard did not play out in such a clear fashion,[7] Emerson stuck to his first principles: Keep fighting men in the field for as many days as possible, move fast and stealthily, and direct the enemy into well-laid traps rather than blunder into random firefights. Each time upon walking into a new area, the American scouts and observers spent days inspecting the terrain to find paths being used by NVA troops in transit. Once the patterns and bivouac locations were known, they radioed back to headquarters. Such a request might read: Lay down a barrage at 0200 hours on the following enemy areas, but not the adjacent friendly ones.

The artillery fire was not to destroy the enemy in camp but to roust him. "So the VC would go running down the trail, just like anybody else would," says Brien Richards, an infantryman for the Second Battalion. "The artillery would keep 'em moving and they never knew where we were." The Recondos were lying in wait inside "friendly" areas, along the expected path of NVA movement. Emerson's men had prepared fields of fire. They used grenade launchers and automatic weapons, but the most devastating weapons were Claymore mines, which, when detonated, sprayed shrapnel like massed machine guns.[8] Taylor says Emerson's 1965 paper came closest to fruition in an action in War Zone D, appropriately called Operation Checkerboard.*

In many variations, the Checkerboard usually worked. According to a combat report in *Newsweek,* by mid-1966 Emerson's battalion was beating the enemy on its home field: moving like ghosts, laying ambushes, and hauling rucksacks with maximum ammunition and minimum rations. A typical ration was a little canned meat and a few pounds of rice carried in a spare sock, which lasted one man three to five days until the next resupply mission by air. "We went out heavily gunned, carrying ammo, ammo, and more ammo, then water," says Richards.

"They stayed totally on the move," Emerson says of his men. "Instead of me picking out an ambush for them based on old Japanese infantry maps, the decisions were decentralized to the lowest level."

---

* In such open country there were no well-defined corridors of foot travel, so the battalion set up a checkerboard of friendly and enemy grid-squares.

Helicopters of the First Brigade kept Emerson's men hustling to NVA footpaths all over central South Vietnam, from Pleiku on the west to Tuy Hoa on the coast to the suburbs of Saigon. What the helicopters didn't do was run the men back to camp for hot food, showers, and beer. According to Jim Gould of the Recondos, the troops stayed in the field for as much as three weeks at a stretch, followed by a three-day break. "Then it's back out we go," Gould says. "In the early days we did not have what was known as fire bases. We just traveled around the country: set up an LZ, ran the mission, and moved on to another LZ."

Emerson extended his tour an extra six months, but the army moved him out of country in October 1966. He began laying plans for a return.

At this time the 101st Airborne was just one of several infantry divisions operating in Vietnam. All had come to rely on helicopters in various measure for mobility, air support, and resupply. A hotbed of helicopter experimentation was the army's First Cavalry Division (Airmobile), which arrived at the port of Quinhon in September 1965. It drew on war games dating back to 1955 in building its operations around helicopter travel to an extent never attempted in previous wars. "Helicopters in Korea were essentially for air evacuation and observation," says Charlie Siler, who was an information officer for the division upon its arrival in Vietnam. "But in the Airmobile Division, aircraft were the primary vehicle for deployment. We did not try to keep landlines open to truck convoys, and that was a radical change."

Making it possible was a new generation of helicopters. In 1963 the army began moving out of Korean-derived piston-powered helicopters to a new generation, all powered by gas turbines. The stalwart was the Bell UH-1 Iroquois, universally known as the Huey. Legendary status notwithstanding, the Huey entered the war with woefully inadequate lifting capacity: With a full four-man crew on board and just one patient, the early medevac Hueys could barely get off the ground when working in the Central Highlands. Performance and cargo capacity improved greatly with the Huey "D" model and the penultimate "H" model. The Huey's versatility and toughness explained why Bell Helicopters built four out of every five helicopters used in Vietnam. But Hueys in flight were noisy—the

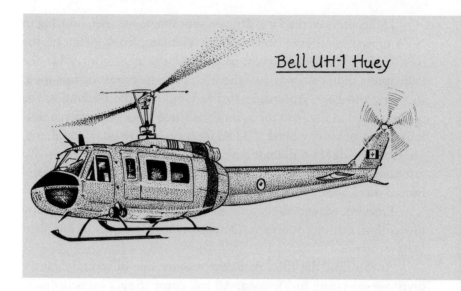

Bell UH-1 Huey

twin blades made a whapping noise that could be heard easily at a distance of two miles, and farther when the enemy dug listening holes in a hillside—so pilots needing to sneak up on the enemy learned to hug the tree cover in a tactic called "nap of the earth flying," popping up at the last moment to prepare for a landing or for a strafing run. Before slicks attempted to land in an disputed area, gunships and fighter-bombers saturated the area with ordnance to push back any enemy and to set off rotorwash-activated "helicopter traps."

Pilots liked the way the Huey's heavy rotor blades stored rotational energy, which allowed them to make well-controlled emergency landings. In the opinion of some pilots, those blades also allowed the Huey to make its own hover hole, employing the main rotor like a blunt machete, bashing through vegetation on the way down to the ground.

Sometimes it worked, if the branches were light. Sometimes it didn't. Either way, cutting a hover hole was a sobering experience. Pilot Robert Steinbrunn, a pilot for the Seventeenth Cavalry Regiment (Air), recalls his attempt to cut a path down to a combat team in Vietnam. While a crew member called out directions to keep the fragile tail rotor in the clear, he lowered his helicopter into a bamboo thicket. "I got halfway down before I lost my nerve," he says. "The bamboo was crashing in around me and the noise and the sight of it were too much." Link Luckett, who flew helicopters in Laos and

Vietnam for Air America, avoided branch-chopping because the impact flattened the leading edge of the blades, reducing lift.

Gene Roberts, Saigon bureau chief of the *New York Times,* saw tree-chopping in action aboard a helicopter bringing supplies into a long-range reconnaissance patrol near the demilitarized zone. The valley was so narrow that the rotor blades were clipping branches and leaves on both sides. "The idea was to get close to a patrol so that the supplies didn't end up in the wrong hands," Roberts says. "Someone with a brick in his hands probably could have brought the helicopter down."

Tracking a large-scale counterinsurgency war was a challenge for correspondents, but helicopters made it possible. "There was no established front," says Roberts. "There were battles in one location and next day they'd be in another, and usually these were separated by long distances. Eighty-five percent of the time, a helicopter was the way to go. Virtually everybody who reported for any period of time used them considerably. . . . As risky as helicopter travel could be, other forms of travel could be far worse."

Helicopters were seemingly as common as jeeps and trucks in Normandy after D-day, so plentiful* that rides were available on short notice. "I'd call the operations desk and tell them I have a mission and give them my destination," says Charlie Siler, a major with the First Cavalry. "Any officer could do that, and sergeants could too if they had orders. No paperwork from me was necessary. It was routine, just like traveling by any other vehicle."

For protocol purposes within the U.S. military (known as MACV, for Military Assistance Command, Vietnam), reporters and photographers held the nominal rank of major in Vietnam, but they didn't necessarily have the same helicopter rights as real officers. Getting a seat could be a matter of personal connections at MACV, timing, or sheer persistence. "You could wait at an airfield for hours before you caught a ride," Eugene Roberts says. Then he met Jurate Kazickas, who came to Vietnam as a freelance correspondent with credentials from several small publications, including a Lithuanian workers' weekly.[9] "I

---

* The use by special forces troops for transportation in the mountains made an impression on even the remotest Montagnard villages. Women of the Arup sub-group were incorporating helicopter images into many textiles by 1965. One remarkable weaving noted by RAND Corporation anthropologist Gerald Hickey depicted villagers waiting to board Chinooks and Hueys for evacuation.

discovered when she was on an airfield," Roberts says, "helicopters would appear out of nowhere, and I could catch a ride with her."

Hitching rides on American helicopters made it possible for photographers to stay in the combat areas for weeks at a time, since they had no reason to return to Saigon or the major military bases as long as their film supplies held out. They sent exposed film back with reporters, with pilots, or with public-affairs officers. Photographer Henri Huet rode helicopters routinely for his work, though he was terrified of them.*

The Huey also doubled in the early years as a gunship, but the external guns added so much aerodynamic drag that it could not keep up with "slicks," the troop-carrying helicopters. In response, Bell Helicopter installed a Huey's mechanical system into a sleek airframe that was dedicated to carrying and aiming weapons. The AH-1 Cobra gunship was the first helicopter manufactured specifically as a weapon mount. It sacrificed all other functions to make room for Gatling guns, rocket pods, and automatic grenade launchers. The two-man crew sat single file—gunner up front, then pilot—which narrowed the profile and made it difficult for enemy gunners to hit as it approached a target.

With time and innovation, the gunship population in Vietnam rose from a total of 20 in 1963 to 700 six years later. At landing zones, gunships and fixed-wing fighter-bombers laid down an advance barrage to protect troop-carrying helicopters. The barrage was to throw off machine gunners from their aim, and to detonate helicopter traps. Much of the firepower did no more than to cut down trees, but some of it was quite effective. After the war, General Benjamin Harrison tracked down and interviewed several North Vietnamese commanders who had tried to overrun his forces at Firebase Ripcord in 1970, in researching a book on the battle. His retired enemies mentioned that the NVA feared American gunships for the way they poured fire directly into bunkers dug into the sides of narrow valleys,** where no other aircraft could touch them—not even the fearsome bombing runs by waves of B-52 bombers.

* In 1971 Huet died in a helicopter after it was shot down during the Lam Son 719 invasion into Laos.

** A contemporary example is America's use of Apache gunships to kill fighters who are lodged in high-rise apartments within sniping range of Baghdad's Green Zone.

Gunships had their role, but were no substitute for work on the ground. That was the job of the infantry and commanders like Colonel Hank Emerson. He returned to Vietnam for his second combat tour in February 1968. He was to command the First Brigade with the Ninth Infantry Division (the Old Reliables) in the Mekong Delta. The principal mission of the brigade seemed simple: to keep a sixteen-mile stretch of Highway 4, and its bridges, open for traffic. This highway connected Saigon with the vital rice-growing regions to the south.

The Delta had little in common with the sparsely populated, heavily forested areas of Emerson's first tour, except for the presence of a die-hard enemy. However successful up north, the Checkerboard couldn't work here because that method required men to walk long distances without being noticed. The open terrain and the hip-deep mud of rice paddies made long treks impossible, and the many villages and their ever-alert dogs would foil any attempt at stealth. And Checkerboard's wide-ranging artillery fire would have threatened the closely packed villages.

The war in the Delta had not been going well. Through 1964 Highway 4 had been safe for traffic, but beginning with the 1965 offensive the Vietcong had frequently been able to blockade it or blow up its bridges. At the time Emerson took charge, helicopter pilots were reporting flocks of little red Vietcong flags across the countryside; once they had been modestly tucked away by day, but not anymore. For a year armed powerboats along the canals, called the Mobile Riverine Force, had been able to break up enemy concentrations, but then the Vietcong had mapped all the navigable canals and learned how to ambush the boats with rocket-propelled grenades and heavy machine guns.

"The psychological situation in the Delta was very bad," the commander of the Ninth Infantry Division, General Julian J. Ewell* told an interviewer later. "I wouldn't say it was defeatist, but it was tense and very nervous. The troops weren't quite sure whether they would make it." Ewell gave Emerson, who was one of three brigade commanders, permission to try something new.

---

* Ewell had long experience in combat. At age twenty-nine he commanded the 501st Parachute Infantry Regiment at Bastogne until injured while firing a bazooka at a German tank.

When Emerson arrived, American infantry tactics in the Delta had mostly been variations on the "sweep," an ancient tactic for hunting wild animals. At its simplest, the sweep had troops walking in lines across the landscape to flush the enemy from hiding. When and if the enemy appeared, the Americans called in artillery or gunships to kill as many as possible before they scattered. Though the sweep sounded intimidating—like hunters closing in on foxes—usually the enemy units remained snug in their bunkers, and if caught outside were savvy enough to slip away via tunnels, by waiting for darkness, or by using canals and ditches. "Although you'd occasionally corner someone and beat up on them, often it was a dry hole," Ewell says now.

"The old idea wasn't working," Emerson says. "The Delta was full of mines, deep mud, and paddy dikes, so we couldn't get around on foot. I said, 'We've got to get airmobile! This is one big LZ!'"

Emerson came up with an approach he called the Jitterbug. As with the high-energy swing dance of the 1930s for which it was named, forces carrying out the Jitterbug maneuver tried to be everywhere at once. The concept was to find large groups of enemy soldiers, pry them out of hiding, and then confuse them once in the open so they couldn't see a way out. It required a high level of activity.

But a Jitterbug operation began long before any shooting, by sifting all available information to divine where an enemy main-force unit might be hiding. Emerson knew the identities of the main-force units in his area, but the location of their hidden heavy weapons and companies at any given time was an ever-changing secret. Learning it fast enough to mobilize a large-scale attack relied on casting a wide net with village informers, newly invented starlight scopes, radar coupled with long-range listening gear, and reports from twelve-man reconnaissance patrols that ranged out each night, when the Vietcong were active.[10] The brigade also experimented with the "people sniffer," a laboratory-in-a-crate that fitted into a Huey and detected the presence of ammonia, which all encampments emitted. The sniffer worked in open and lightly forested areas, but only until the enemy adopted some simple steps to frustrate the equipment.*

---

* One effective countermeasure was hanging buckets of urine from tree branches.

Once the enemy position was identified with reasonable probability, helicopters moved in and shifted troops at a rapid pace. This
frenetic activity was intended to alarm the enemy and persuade them
that they were nearly surrounded, which prompted them to bring
out 12.7-millimeter machine guns to use against the helicopters. The
distinctive green tracer rounds (which looked as big as balloons as
they came up to meet helicopter crews) pinpointed the guns' locations and also indicated the value of the force. The more antiaircraft
(AA) fire the better: heavy fire indicated a main-force unit, and
Emerson was out to destroy the biggest forces he could find. If
Emerson was satisfied that the enemy was within the Recondos'
grasp, he transmitted the radio signal "Pile on!" which meant that all
infantry units were to hike to the nearest helicopter pickup zone and
move in to create a tight perimeter around the enemy.

The goal was to get the Vietcong out of shelter and trapped inside a ring about one half mile across. To block exits via the paddy
canals, troops draped concertina wire across the waterways and
tossed grenades in the water at frequent intervals to discourage escape. Locating the enemy and creating a tight seal sometimes required seventy-two hours of unceasing activity. When the ring was
secure, helicopters dropped bundles of tear-gas bombs, particularly
along lines of nipapalm trees where bunkers were likely to be con-
cealed. Pilots hovered their aircraft among the nipapalms to spread
the agent for maximum effect.

Anyone caught in the ring came under a blanket of fire delivered simultaneously by fixed-wing aircraft, helicopter gunships, and
artillery. "We had great artillery officers, like Joe Wallace and Bob
Dirmeyer," Emerson says.[11] "Dirmeyer could figure out how to deliver what I needed and deal it out all at once. We worked out aircraft
flight patterns so they wouldn't get hit by artillery."* Called "shooting the doughnut," Dirmeyer's intent was to put every howitzer
round within the area ringed by American troops, and to keep doing
this even as the Americans shrank the circle by closing in. At intervals
the firing would stop so that bullhorns could transmit a call for surrender.

---

* Even so, the danger to aircraft was ever present. Brigade commander Colonel Dale Crittenberger of
the Ninth Infantry Division died in a helicopter collision during such an operation.

"I hate buzzwords but in this case synergy was the right word—each type of fire protecting the other—and suppressing the AA," Emerson says. "It was synergistic as hell."

According to a dispatch by Associated Press reporter Peter Arnett—who was as skeptical of military quackery as any correspondent in Vietnam—by early 1969 the Ninth Infantry Division was a formidable machine of war, even in the northern Mekong Delta, the Vietcong's once-fearsome stronghold. It was an amazing performance considering that as recently as 1965, the Vietcong main-force units of the Delta were regarded as unbeatable. Arnett did express doubts about whether America's firepower was leading anywhere productive, but wrote that when it came to dealing out carnage with the "pile-on" tactic, the draftees of the Old Reliables were masters.

"The game of hide and seek has honed the Ninth Division into a deadly efficient machine," Arnett wrote for the Associated Press in April 1969. "One recently captured document from the enemy high command described it as the most dangerous division in the country." Because the style of such warfare required close-in leadership from all officers, the toll was high. "There were six battalion commanders when I was with First Brigade. Three were killed," recalls one of Emerson's battalion commanders, James Lindsay, now a retired four-star general.*

"Using this technique in Dinh Tuang Province, Colonel Emerson, over a period of months, essentially, was able to break up every Communist battalion in the province," Ewell said in an interview with military historians. Emerson's men later repeated the achievement in the notoriously dangerous Long An Province.** "The only problem with the Jitterbug was that it was so complex that it took a real master to do it well," Ewell said.

"Jitterbug was most practical with enough helicopters to give combat units a high degree of mobility," Julian Ewell says now. "Hank's virtue was bringing tremendous intensity to the mission. He was a fighter and combined a high level of tenacity and speed with the Jitterbug. Hank's sole purpose in coming to Vietnam was to beat the

---

* These were Lieutenant Colonels Fred Van Deusen, Bill Bergenac, and Don Schroeder.

** MACV estimated that Vietcong forces controlled 75 percent of the villages in Long An as of January 1964, despite the Americans' use of fortified "strategic hamlets."

living stew out of the VC with minimum American casualties." (For his part, Emerson is emphatic about crediting his battalion officers and troops for the victories. "A lot of people made me look good," he says. Ewell allowed Emerson to handpick his combat commanders.)

"Emerson had a great deal of initiative—higher-ups didn't always like it but he just said screw you, and he set the rules," Brien Richards, veteran of the 101st Airborne Division's early Recondo missions, says. One of Richards's proudest possessions is a signed photo from the Gunfighter.

In the photo, Emerson is wearing a Colt .45-caliber single-action revolver, which in April 1968 he used to shoot two Vietcong soldiers on the bank of a canal. That gun battle and his other actions that day won him a Distinguished Service Cross, but Emerson is almost apologetic: "It was a happenstance, of course—I'm not that stupid—but I thought I'd cut 'em off. . . . It was a rash thing to do."

"As good as a man he was, he had some quirks," Richards says. Emerson acknowledges that on occasion he carried Scotch whiskey in his canteens for medicinal purposes.

The Gunfighter's second tour in Vietnam ended with a bang on August 26, 1968, when a Vietcong gunner launched a rocket-propelled grenade at his helicopter as he was returning to headquarters from a low-level reconnaissance mission. Such projectiles are unguided and the great majority of them miss the aircraft at which they are aimed, but this one blew off the tail boom. The Huey nosed down, started spinning, and dropped into a rice paddy. It rolled over and caught on fire. Two men died instantly; four men climbed out; that left one man, who was trapped in the flames. That was Emerson. With a helping hand from Colonel Ira Hunt, Emerson pulled himself loose shortly before the aircraft erupted in a fireball, which followed from the fact that it had landed on an enemy ammunition dump. He was evacuated in a command and control helicopter and treated for burns and a massive infection at a field hospital. General Creighton Abrams smuggled in a bottle of whiskey and, joined by fellow colonel Ben Harrison, Emerson emptied the fifth in two hours.

Motion pictures often portray the genuine risks that infantry soldiers faced when landing in a "hot LZ," but helicopter travel in Vietnam posed risks to all participants, military and civilian, particularly when Vietcong antiaircraft weapons improved. Statistically,

helicopters were the biggest single killer of the highest-ranking officers: helicopter crashes killed four of eight generals who died in the
war,* along with one admiral. John Paul Vann, an early critic of the
war and the subject of the biography *Bright Shining Lie,* died in a helicopter crash in June 1972.

Emerson recovered from his crash, but never got back to the
war. He ran twelve Jitterbug missions before being shot down for his
third and last time. In 1969, while training at Fort Rucker to qualify
as a helicopter pilot in preparation for a return to Vietnam with the
First Cavalry, Emerson was pulled out of helicopter school one week
before graduation and reassigned to the Eighty-second Airborne as
commander of special forces. Perhaps stateside duty was the best
thing for the unreconstructed warrior, because in 1969 the American
war on Vietnam started winding down.

---

* These were Generals Bruno Hochmuth, George W. Casey, John Dillard, and Keith Ware, the
recipient of a Congressional Medal of Honor. Emerson called Ware "a hell of a stud," his highest
accolade for a fighting man.

Chapter Eleven
Last Man Out

Using helicopters for last-minute evacuation had been rare during wars of the previous decade that featured the aircraft, in Korea, Malaya, French Indochina, and Algeria. Helicopters were for bringing reinforcements and ammunition to the battlefield, and for taking wounded away. In Vietnam, helicopters took on a new role as getaway vehicles, most famously during the final evacuation of Americans and selected allies from Saigon in the final days. However prominent it would be in 1975, though, this role began emerging much earlier in Vietnam.* One trendsetting incident came in June 1965, when marine helicopters evacuated the entire population in and around Khe Tre[1] in the face of oncoming Vietcong forces. Helicopters grabbed up 2,600 people and all their belongings, down to dogs, bags of rice, and bales of tobacco.

About this time, a new market for helicopter rescue was developing on the other side of the

---

* The first use of aircraft for wartime evacuation occurred during the siege of Paris in the Franco-Prussian War. During 1870 and 1871, wealthy citizens flew out of the city via balloon, along with mail. The total number of passengers fleeing the city was approximately one hundred.

demilitarized zone, deep inside North Vietnam and along the northern reaches of the Ho Chi Minh Trail. The air force and the navy began making bombing runs over North Vietnam, and downed pilots
began calling for help.

Any pilot or crewman dropping deep into the territory of a
very angry enemy faced death or a long and brutal captivity. Their
hopes rode with the air force's Thirty-eighth Aerospace Rescue and
Recovery Service, known as the Jolly Greens because of the hulking
look of their Sikorsky HH-3 rescue helicopter. The Jolly Greens did
their work without the usual gunnery on board but had backup from
Douglas A-1 Skyraiders, a propeller-driven airplane well suited for
strafing.

One of the Jolly Greens' most highly decorated pilots was
Captain Leland Kennedy, who, on October 5, 1966, was to receive
the call to duty. It would win him an Air Force Cross, the service's
second-highest decoration. That day an F-4C Phantom (call sign:
Tempest Three) was flying west of Hanoi, escorting an electronic-
warfare aircraft back to base, when a MiG fired a heat-seeking missile
from behind. The missile flew up the tailpipe and destroyed the engine, but without exploding.

"This was over a really bad area," recalls Ed Garland, who occupied the rear seat that day. "There were two rivers on the northwest,
the Red and Black Rivers. If you were north of the Red, the standing
order was: 'No pickups there—it's too dangerous.'" Nursing their
airplane as far south as possible to aid any rescuers, the two pilots
waited till they saw the Red River, then ejected. Lieutenant Garland's
chute dropped him on a mountainside that was forty-five miles south
of Hanoi. That put him 300 miles from his fighter base. He had a pis-
tol, a short-range radio, a bleeding ankle, and a back injury. A jet
passed low overhead and began strafing the ground nearby; it was
an F-105 Thunderchief pilot* who had by a rare chance seen the
Phantom's smoke trail, had watched the parachutes descend, and had
called for rescue. The message was relayed to Udorn Air Force Base.

Garland had seen on his way down that the mountain had a
radar station on its peak. While two Jolly Green helicopters lifted off
from a forward staging area along the Laos border, Garland heard

---

* This pilot never identified himself and Garland never learned his name.

soldiers from that station coming down the slope and closing in, despite the Thunderchief's strafing. Though he would be on the ground for a full four hours, the enemy held back from capturing him. (Garland learned the reason later: He was bait for a trap. The North Vietnamese had been busy setting up heavy machine guns for the helicopters they knew would be coming.) Garland received a final radio message from his counterpart, Bill Andrews, on the other side of the mountain: Andrews had been hit in the chest, he was losing consciousness, and he was going to shoot up his radio to prevent the enemy from luring rescuers with it.[2]

When Kennedy's helicopter (Jolly Green Two) left its forward base,[3] it was supposed to serve only as escort, known as the "high bird" position, while Captain Oliver O'Mara and crew in the primary helicopter, the "low bird," carried out the rescue.

When the two helicopters arrived, says Garland, "the low bird came in and it drew small-arms fire from rifles and machine guns, anything they had." By O'Mara's third attempt to get a rescue seat to Garland on the mountainside, enemy gunfire had destroyed the helicopter's hoist, shot up the fuselage, and damaged one engine. Tagged out, O'Mara left.

Kennedy was flying high bird that day because he was still a novice in the rescue business. It was his eighth mission out of Udorn, but this would be his first actual rescue attempt. The unfavorable circumstances were looking similar to those of a rescue attempt in September 1965 by another helicopter from the Thirty-eighth, call sign Duchy 41. As Duchy 41 hovered in a box canyon, heavy machine-gun fire from above had shot it down. All five crewmen had been captured by Communist forces.

Compounding the difficulty for Kennedy's Jolly Green Two, the four Skyraiders that had been strafing vigorously to hold back the enemy started running out of ammunition. This left Kennedy's helicopter with a single advantage: a fogbank was moving in. Kennedy started his first run, using the fog as cover from the heavy guns on the high ground.

A grenade or missile exploded in the cabin, stunning pararescueman Chief Master Sargeant Ed Williamson. Williamson called out that he was fine and needed help only getting soot out of his eyes. Kennedy backed away and waited for Williamson to regain his sight.

The Skyraiders radioed that they were dry on ammunition but would make low passes to frighten the enemy.

Kennedy made his second run: The fusillade from the ground punched more holes in the helicopter and put a bullet in Williamson's knee. Kennedy retreated again and polled the crew: that had been the Jolly Greens' fifth try. Did they want to cheat death once more? They crew voted in the affirmative. This time Kennedy lowered the jungle penetrator to the ground on its cable and dragged it toward Garland as if trolling for a bottom-dwelling fish.

"I could see the helo now," Garland recalls. "He didn't have much cable out, maybe fifty feet. All the guns were firing at him. I ran maybe thirty or forty feet and pulled the seat down."

Kennedy snatched him off the ground and rose toward the protective fog. There was one final thrill: Aboard the Jolly Green, Williamson looked down with horror at the sight of Garland's parachute rising in tandem with the airman. Somehow the shrouds or harness had tangled with the rescue seat. As soon as the helicopter accelerated, the parachute canopy was going to sweep back and tangle with the tail rotor, which would send the helicopter spinning out of control. Perhaps these men were linked by telepathy, as only saved and savior can be, or perhaps not: in any case, Garland realized the problem instantly and yanked the parachute loose.

His helicopter low on fuel, Kennedy headed back to his Lima site. He won an Air Force Cross again two weeks later, for another rescue in North Vietnam. He was the first airman to receive it twice.

Bravery under fire notwithstanding, the United States began edging toward the exits in April 1969. That was when monthly MACV troop levels began a long decline to official departure in March 1973.

In 1969 the incoming administration of President Richard M. Nixon wanted to preserve public support for the war long enough to secure a peace treaty for the South Vietnamese government. That required a drop in the casualty rate, so American infantry commanders were told to shift war-making duties to ARVN forces. This was called "Vietnamizing" the Vietnam War. The slow diffusion of American units out of the combat zones, it was thought, would not interrupt the steady pressure on the North Vietnamese to accept a permanent partition of Vietnam.

By April 1970 Vietnamization of the war was well under way. One spot of high ground caught between MACV's old offense and its new defense was Fire Support Base Ripcord, an American artillery base operating atop a steep-sided hill in the northwest quadrant of South Vietnam. It was a remote and roadless area, twelve miles from Loas, and was too small for an aircraft landing strip. Helicopters were the only way in, and would be the only way out. The fight for Ripcord would be the last major battle fought by U.S. infantry in Vietnam.

But that would happen in July, and through June 1970 Ripcord still appeared to be a viable strongpoint for carrying out Operation Texas Star, a series of patrols and forward artillery bases to block the movement of North Vietnamese supplies from hidden storehouses in the A Shau Valley. Something like a vast warehouse at the end of a rail line, the A Shau received and stored supplies via a major branch of the Ho Chi Minh Trail. American forces had given up on trying to take and hold the A Shau itself.* But perhaps the flow of supplies from the A Shau, east to enemy forces on the coast, could be choked off without a frontal assault; at any rate, this was the aim of Operation Texas Star. After Texas Star was well under way, the Third Brigade, 101st Airborne Division, would turn the fighting over to ARVN troops in the spirit of Vietnamization.

It would never happen. Ripcord had about 200 troops of one battalion on hand on July 1, 1970, when nine battalions of North Vietnamese began carrying out orders to wipe it out, along with every perimeter camp held by American patrols. The NVA plan was to reduce the base's defenses with barrages of heavy mortar fire and nighttime sapper attacks. Antiaircraft guns dug into nearby hillsides would shoot down helicopters offering assistance. These hillsides were so steep that the air force would not be able to use B-52 bombing raids to dislodge the attackers. Then the NVA would overrun the hilltop. This would remove the threat to the supply lines running out of the A Shau Valley and probably hasten the Americans' exit from Vietnam. The overall plan was much the same as the one that had annihilated French forces at Dien Bien Phu in

---

* An assault on the A Shau in April 1968 called Operation Delaware resulted in ten helicopters shot down in a single day, and twenty-three damaged.

early 1954,* though smaller in scale. Ripcord was even more isolated than Dien Bien Phu.

NVA troops had prepared by digging log-reinforced bunkers into the steep hillsides. Deep tunnels connected many bunkers, so North Vietnamese troops could move to the best firing points without getting hit. While still vulnerable to helicopter gunships that could fire their rockets and miniguns directly into bunkers and slit trenches, well-placed antiaircraft guns and the frequent low cloud cover would dampen this threat also.

The battle began on July 1. After two weeks, American forces at Ripcord were holding on but wearing down. The turning point followed a spectacular mishap on July 18, when Vietnamese machine-gun fire from across the valley brought down a Chinook helicopter that was hovering over an ammunition dump. The helicopter crashed on top of its sling load of ammunition, caught fire, and triggered the entire cache in a spectacular series of thunderous blasts, fuel-air fireballs, and smoke trails. The mayhem lasted eight hours. While still under attack from NVA gunners outside, the troops had to dodge fire from inside as well: from their own cluster bombs, white-phosphorus rounds, grenades, artillery shells, and clouds of tear gas. The chaos also destroyed a battery of 105-millimeter howitzers.

On July 22 the word came: Division headquarters had decided to empty out the base with dozens of helicopters, evacuating equipment as well as men. General Sidney Berry, acting commander of the 101st Airborne in nearby Camp Eagle, had some misgivings when he wrote his wife that morning: "The mountains seem loaded with 12.7mm AA machine guns. Yesterday, we had two more helicopters shot down."

Following an overnight bombardment of the base perimeter intended to discourage the enemy gunners, fourteen CH-47 Chinook helicopters took flight before dawn to begin hoisting everything of military value, including the surviving artillery tubes, radar, and two bulldozers. Evacuation of troops began at 8:30 A.M. Mortar fire in-

---

* Dien Bien Phu was an obscure village 175 miles west of Hanoi, near the Laotian border, when the French decided in late 1953 to dispatch paratroopers to open up a stronghold there. General Vo Nguyen Giap identified the destruction of the base as his best opportunity to break the enemy's will. Using diversionary attacks around the country to disperse the understrength French forces, Giap's Viet Minh forces cut the supply lines to Dien Bien Phu and waited for monsoons that would shut down French air support. Giap attacked in force in March 1954, broached the defenses by digging trenches and tunnels, and launched a final human-wave attack that prompted a mass surrender.

creased as the North Vietnamese realized that all the Americans were leaving. Orders directed them to overrun the base before the evacuation could be completed.

Using a nearby waterfall as a reference point, Huey helicopters traveled a specific approach path, landed, and paused amid the exploding mortar rounds for excruciating seconds to let the troops board. The fire from the hilltops was so intense at first that troops were reluctant to leave the safety of bunkers and trenches to get on the Hueys. But they decided that things weren't going to get any better. Once loaded, the helicopters turned and dived off the hilltop. The evacuation was over shortly after noon, at a high cost to the helicopter fleet* but with surprisingly few human casualties.

Over the five years of U.S. occupation remaining after the demise of Firebase Ripcord, helicopters found other work in and around Vietnam, and were transferred in squadron quantities to the South Vietnamese Air Force.

The last major contribution to a battle by American-piloted helicopters in Vietnam came during an invasion from North Vietnam in March 1972, when the NVA drove long columns of Soviet-built tanks into South Vietnam.⁴ One objective of the Easter Offensive was the destruction of South Vietnam's Fifth Division, sixty miles north of Saigon. The tanks entered from Cambodia, and formed a ring around the town of An Loc. Weeks went by and no help got through. Even elite troops were stalled ten miles short.

Meanwhile a test of a new tank-killing TOW missile** and its aiming system was under way at Fort Lewis, Washington, with no immediate plans to use it. The helicopter intended to carry the TOW, the AH-56A, wasn't ready yet, so it was mounted on two obsolete Huey Model B helicopters owned by the contractor, Hughes Aircraft. But the Easter Offensive wasn't going well and the army decided in a flash that this was an excellent time to try it against Soviet tanks invading Vietnam, though the team had not yet fired an

* Of twenty Hueys used in the operation by the Ghostrider Company of the 158th Aviation Battalion, eleven were destroyed or damaged.

** TOW was short for "Tube-launched, Optically tracked, Wire-guided missile."

armed TOW missile against anything. The army ordered the entire package—civilian helicopters, civilian technicians, and civilian instruments—to Vietnam and gave them one week to go into battle. They were in the field by April 26. The first TOW destroyed an enemy tank near Kontum on May 2 with a single shot, then destroyed three more tanks the same morning. In one month, TOW missiles destroyed twenty-four enemy tanks from the air. While the TOW was not the first precision-guided missile fired from helicopters—that was the French AS-11—it was deadlier.

Even after the negotiation of a second peace agreement in March 1973, American helicopters found some work in the next two years. In the effort called Operation Endsweep, Sea Stallion helicopters helped remove thousands of sea mines from Haiphong Harbor in North Vietnam, where they had been dropped by other navy aircraft just ten months before.

And American troops in helicopters continued flying over the countryside that year, but now they were unarmed; it was part of an effort to locate and return soldiers' remains to the United States. Unreconstructed Vietcong shot down one such search helicopter in December 1973.

Despite all the goodwill gestures by departing Americans, and despite the peace treaties signed in Paris, the North Vietnamese had every intention of taking control of the entire country. Mindful of the possibility that the United States might send its troops back if provoked, though, the North Vietnamese Army delayed a full-scale invasion of the South until it could restore its fighting strength following heavy losses in 1971 and 1972.

NVA forces began their final assault on South Vietnam in January 1975, beginning with an attack in Phuoc Long Province in the far north, near the Cambodian border. Resistance was lighter than expected and two months later the NVA advanced on the centrally located, and strategically vital, city of Ban Me Thuot.

After three NVA divisions stormed Ban Me Thuot, President Nguyen Van Thieu issued orders to his commanders to abandon the Central Highlands. Refugees jammed Route 7B leading east to the ocean. Thousands of ARVN troops, many of whom had dumped their weapons and were trying to flee in the same direction, decided that they should be looking after their families because chaos was about to swamp them all. The city of Hue fell, and Da Nang, and air bases farther down the coast. After realizing that three North Vietnamese divisions were certain to take the air base at Nha Trang, commanders of the South Vietnam's Third Airborne Brigade headed for the airport, climbed aboard their command helicopters, and made a clean getaway.* The end was in sight. By the end of March, hundreds of thousands of Vietnamese were leaving, or preparing to leave, the country by airplane and boat.

The fall of Saigon at the end of April saw a historical peak of helicopters' employment as getaway vehicle. At least eighty helicopters, some carrying five times as many people as their rated capacity, played an unexpectedly important role during the last twenty hours of the U.S. presence, because the U.S. ambassador Graham Martin had been resisting all suggestions to authorize his citizens' departure. Martin had even blocked the CIA from pursuing some of its own plans to set up emergency helipads and fuel caches on rooftops, fearing that any signs of weakening U.S. resolve would set off a panic.

Martin hoped until the very end that something would halt the North Vietnamese short of Saigon, leaving it an "open city." Perhaps the ARVN forces would rally and stop the fast-moving columns of North Vietnamese coming toward Saigon on Route 1 after their blitzkrieg down the coast of the South China Sea. For his part, President Van Thieu speculated that the United States might turn their B-52 bombers loose on the advancing army. But Martin's hope to shield thousands of loyal South Vietnamese in Saigon looked more like folly than fealty after April 20, when the last bastion of ARVN resistance fell at Xuan Loc.

President Gerald Ford told the press that Vietnam was lost. Had B-52 bombers slowed the advance, it would have mattered little since the South Vietnamese forces were in disarray and thousands of NVA troops were already dug in around Saigon, having infiltrated the suburbs months earlier. Meanwhile, enemy units were working to cut a link essential to the original escape plan, which was to have thousands of loyal Vietnamese gather at the Defense Attache's Office, a yellow building at Tan Son Nhut Air Base, and fly out on the largest

* While it is tempting to criticize the use of helicopters as escape pods for ARVN's power elite, it is now standard practice. Beginning in 1957, the air force provided two Bell 47J helicopters to President Eisenhower and his family, chiefly to enable their prompt evacuation in the case of nuclear attack.

transport aircraft available. That base came under attack on April 28 from NVA rocket batteries and renegade South Vietnamese pilots.

On the morning of April 29, President Ford ended the agony that Martin's hesitation had created. He ordered immediate evacuation of hundreds of American and European civilians, marine guards, and approximately 7,000 South Vietnamese whose names appeared on an approved list. They were to fly to carriers of the Seventh Fleet, forty minutes by air from Saigon.

The message that Operation Frequent Wind was actually happening went out by telephone and as a code signaled by playing "White Christmas" on American Radio Services. By early afternoon city-wide hysteria was under way, what Sigmund Freud once called the "gigantic and senseless fear" of a city under attack.

United Press International photographer Hubert van Es took the photo that became the icon for America's last days in Vietnam: It shows a line of people climbing a ladder to a small building on a roof. Perched on that building is a Huey with civilian markings. Many newspapers (and later the musical *Miss Saigon*) mislabeled the location as the roof of the American embassy in Saigon.

In fact the photograph was of the Pittman Apartments on Gia Long Street, and the cubical building was an equipment shed for the building's elevator. The Pittman was a main residence for employees of the CIA and the U.S. Agency for International Development. Most of the evacuees on the Pittman's roof were South Vietnamese government officials, who had gathered there after CIA station chief Thomas Polgar sent them word. The helicopter was operated by Air America, a CIA-backed airline. In all, twenty-eight Air America helicopter crews shuttled people from improvised helipads in and around Saigon. Since the Hueys could fit into small landing zones, they were a significant addition to the force of larger aircraft officially assigned to the job: ten air force helicopters and two squadrons of marine helicopters.

That night the center of evacuees' attention shifted from Air America's helipads and Tan Son Nhut to the U.S. embassy. The embassy was a half mile from the Pittman Apartments, from which the last of the CIA employees had already departed. It was from the embassy that the last helicopter flights, flown by marine and air force crews, would leave Vietnam, flying directly to ships of the Seventh Fleet. After a tree was felled, the big Sea Stallion helicopters headed

CH-46 Sea Knight

for a helipad on a parking lot inside the walled compound; the smaller Sea Knights landed on the rooftop helipad, a spot known to staff as Bunker's Hill after the previous ambassador, Ellsworth Bunker. Parents hurled their babies over the walls.

In the frenzy, any Vietnamese who fought his way to the front of the mob was likely to get a helicopter ride, regardless of whether his name was on a list. There was no opportunity to double-check credentials. Guards joked that they were dispatching North Vietnamese corporals to Guam. Just before 5:00 A.M., the order came from Washington: The helicopters were to load only Americans from now on.

Graham Martin remained in the embassy until 5:30 A.M., clinging to the hope of arranging escape for few hundred more Vietnamese. He gave it up and left, clutching a folded American flag, after receiving direct orders from President Ford. Somehow the word spread to Vietnamese waiting outside the embassy that the last Americans were about to leave. So many Vietnamese surged over the walls that the marine guards retreated into the embassy, surrendering one floor at a time as they moved upward to the last helipad in the compound, on the roof. For over two hours the marines defended the steel door to the helipad with grenades, tear gas, and rifles, wondering all the while whether one last helicopter would bring them off. Shortly before 8:00 A.M. the last helicopter out of Saigon, a Sea Knight, arrived. As they lifted off, the eleven Americans looked down at hundreds of Vietnamese still waiting for a few more helicopters. They never came; in Washington, the presidential secretary announced that the last helicopter had taken to the air. A North Vietnamese tank broke down the embassy gates three hours later.

Out at sea, among fifteen ships of the Seventh Fleet, the scene was only slightly less wild. Joining the stream of helicopters from the embassy were at least thirty helicopters flown by South Vietnamese pilots on their own initiative. Seven helicopters gathered at the helipad aboard the USS *Blue Ridge;* one landed atop another and almost tipped into the ocean. The jam on the carrier decks was so extreme that some pilots had to ditch their aircraft at sea and await rescue; other helicopters were allowed to land but after the last passengers stepped onto the ship, sailors tipped them over the side to make room.

The image of turbine-powered helicopters flopping into the South China Sea, as if equivalent to bicycles or wheelbarrows too rusty for repair, was emblematic of this profligate war's ending. Of the helicopters that the Americans brought into Vietnam, almost 5,000 crashed due to mishap or hostile fire. They were flown by 40,000 pilots with the U.S. armed forces, the Vietnamese Air Force, and civilian contractors. The toll was 2,197 pilots and 2,718 crewmembers killed. Still, despite all the losses and the many articles stateside about crashes, helicopters proved durable given the heavy use. The loss rate was one fatal crash per 20,000 missions.

In any case, 12,000 helicopters were not enough to win the war. "It's been said that we didn't lose battles in Vietnam," reflects Hank Emerson. "The truth is that we lost a lot of battles and it was covered up. Once the Vietnam War was over, we couldn't forget it fast enough."

Among the lessons from the cauldron could be one regarding roles for helicopters when first plotting a counterinsurgency war. Perhaps helicopters should be at the back of commanders' minds, rather than at the front. That is, commanders should first sketch out a plan that is based on the best information available about the enemy's territory, industry, culture, history, and supply lines. How will critical territory be taken and held? Only after this big picture is clear should helicopters be considered, along with other assets. In some engagements helicopters might play only a guarded, minimal role (as in Checkerboard). In others (as in Jitterbug) they may be vital to success. Victory happens on the ground, not in the air.

Chapter Twelve
Back in the World

More American helicopters survived Vietnam than were shot down or lost to mishap. One was a Huey in storage at the former U.S. base at Can Tho, abandoned like hundreds of others to a victor who had barely used helicopters at all during the conflict.* A year after the war, former ARVN pilot Ho Kan Hai, with his wife and four children, stole it from the base. It still had U.S. markings when he landed in Thailand.

In the ensuing years, the Vietnamese sold their unneeded helicopters and spare parts for whatever the world market would pay, mostly through third parties, such as Thailand. By contrast, the U. S. government and the rotorcraft industry wanted to avoid flooding the civilian market with thousands of war-weary helicopters. Could enough new uses be found? One sign that they could was that sales of new helicopters for commercial uses rose threefold from 1970 to 1975. While helicopters would never be available in such profusion as during Vietnam, they spread

---

* North Vietnamese government officials used Soviet-piloted helicopters on occasion, but there were few sightings of NVA military helicopters during the war.

rapidly into new niches. When compared to the tiny number of commercial helicopters in use shortly before Vietnam—less than four hundred—the contrast is striking.

In June 1976, three years after the Mideast oil embargo and the new era of sharply higher crude-oil prices, the Saudi Arabian kingdom had billions of petro-dollars to spend. The money was flowing out easily enough from the royal coffers to vendors worldwide, but goods weren't getting back in. This was obvious from any viewing angle in the 1,400-year-old port of Jeddah, on the country's eastern Red Sea coast. A half-million tons of goods, some in crates, some in bags, and some fully exposed to the sun and infrequent rains, lay in heaps across eight square miles. Broken bags spilled flour across sandy lots where it caked in the rains and mixed with the sand. Sickly cows in makeshift corrals staggered to stay upright. Thousands of dusty new cars sat door-to-door in parking lots, their glassed interiors reaching hellish temperatures under the sun. It was "an extraordinary mix of opulence and squalor," according to Edward R. F. Sheehan of the *New York Times*.

The situation was worse offshore, where two hundred ships lay at anchor with another million tons to unload. Some vessels had waited a turn at the wharf for months, where Yemeni and Egyptian stevedores could use dockside cranes to unload. Ship owners put up with the harbor crisis because the Saudis were willing to pay "demurrage" penalties of thousands of dollars for every day of delay.

Considering the range of goods trapped in the traffic jam, the most valuable of which included X-ray machines and Rolls-Royces, it might seem that hundreds of thousands of bags of powdery cement in cargo holds would be at the bottom of the unloading priority list. But the construction of buildings across the kingdom depended on cement, and many of these cement-holding freighters lay more than a mile from the docks. While flat-topped barges could get the cement bags closer to shore, the docks were so backed up that the stevedores couldn't unload the barges either. Frank Carson of Carson Helicopters of Perkasie, Pennsylvania, flew to Saudi Arabia to propose a solution. Carson specialized in the lifting of heavy cargoes on lines slung from underneath helicopters; pilots call it "long-line" work. After weeks of office visits, Carson convinced the interior ministry that a small fleet of helicopters and thirty pilots could get the cement moving. The helicopter was the Sikorsky S-58T Twinpac, a Choctaw given new life by the replacement of the gasoline-powered radial engine with two turbine engines. Many of the pilots and mechanics were from Air America, the CIA-backed transport operation working in Laos, Cambodia, and Vietnam.

The cement lift began in June. Each helicopter carried an electrically operated hook at the end of a hundred-foot steel cable. Every ninety seconds a pilot would repeat this trip: Pull up over a cement barge and place his cargo hook directly into the hands of a rigger, who would clip it to a sturdy net holding forty-four hundred-pound bags; the pilot would fly the load of bags a mile over the water to the pavement outside a warehouse and set it down, stirring up a cloud of cement dust as he did so; the pilot would open the hook electrically; and then he would zoom off for another load. Four helicopters flew the loop continuously and any mistakes threw off everyone's pace, so the company expected pilots to fly their four-hour shifts with no errors and no wasted motions. Pilots flew fifty-six days without a break, then had four weeks off. Mechanics worked however many hours it took to deal with high engine temperatures, heavy use, and cement dust that was in the air continuously. Each night dew fell, and mornings revealed helicopters covered with a gray, plaster-like shell that had to be knocked loose.

Flying net-loads of bags at the rate of two tons per trip, making a thousand round trips every day between barges and the warehouse, the Carson helicopters moved 660,000 tons of concrete in one year. In 15,000 hours of flying, the operation suffered three helicopter crashes, one fatal.

Another job for Vietnam-veteran pilots in the late 1970s was instructing new pilots in Isfahan, Iran, for Bell Helicopter International because Shah Reza Pahlevi had been directing millions of that nation's oil dollars into helicopters. Among them was a fleet of two hundred AH-1J Sea Cobra gunships, purchased beginning in 1975. The Bell training program ended abruptly with the Islamic revolution of 1979 and the instructors dispersed, but now Iran had in place a cadre of very capable gunships and pilots, along with sufficient ammunition.

Iraqi leader Saddam Hussein watched the chaos of that revolution and launched an attack on Iran in 1980 with the intent of taking key territory along the border, figuring that the Iranians would be too distracted and divided to fight back. The war began when Iraqi air forces attacked nine Iranian air bases on September 22. After less than two weeks, Iraqi forces had taken portions of Khuzestan and territory along the Shatt al Arab waterway, the two main objectives for the Iraqi invasion. At the end of September, military observers could only speculate about why the Iranians hadn't used any of their Sea Cobra helicopters, some of which carried TOW antitank missiles. These were the same make of helicopter-launched missiles used so effectively against North Vietnamese tanks in 1972.

The reason that Sea Cobras hadn't appeared earlier was that hundreds of fixed-wing and helicopter pilots were in jail, following an attempted coup by Iranian pilots in July 1980. The new regime regarded all Western-trained aviators as tainted. But now six divisions of Iraqis were headed for areas considered essential to Iran, including two big pipelines providing oil and natural gas to Tehran. As an Iraqi armored division approached Dezful Air Base on October 4, Iranian president Bani Sadr ordered the release of jailed pilots. The Sea Cobras were soon in action and the Iraqis never took the city of Dezful.

The Iranians used their Sea Cobras repeatedly during the war against Iraqi tanks and occasionally in dogfights with Iraqi helicopters, at the cost of half their fleet. In one instance, a single Iranian gunship held up an armored brigade. Later, Sea Cobra losses rose sharply after the Iraqi armored crews learned how to use the 12.7-millimeter machine guns on their tanks. DShK heavy machine guns, carried by mule and horse, were equally deadly to Soviet helicopter transports and gunships in Afghanistan.*

Halfway through the Iran-Iraq War, a helicopter-based struggle of a different sort was shaping up in the Karakoram Range. Starting in 1984, it was the highest-altitude war in history.

---

* Notwithstanding much publicity about the Stinger missile that the CIA began providing to Mujahadeen fighters in late 1986, the Soviet Air Force actually lost helicopters at a faster rate in the two years before the Stinger arrived, principally to ambushes using 12.7-millimeter and 14.5-millimeter guns.

The dispute dated to 1949 and a disagreement over the exact course of the India-Pakistan border where it passed through the old kingdom of Kashmir. The disagreement was academic until an Indian army officer noticed in 1977 that the Pakistanis were issuing permits for mountaineering parties to climb certain high mountains that India claimed. A race was on to control the Siachen Glacier[1] and three high passes.

In a secret mission called Operation Cloud Messenger, the Indian Army used helicopters to reach the high ground first, in April 1984. Indian troops planted fiberglass igloos at altitudes as high as 22,000 feet in the Saltoro Range forming the west rim of the glacier. Most of the fighting was conducted with cannons and mortars, which fired whenever the weather was clear enough to pick out a target. Indian Mi-8 helicopters brought light cannons to 17,000 feet and troops dragged the hardware the rest of the way, a few agonizing feet at a time. While the lower-altitude Pakistanis could depend on trucks and pack animals, Indian forces were totally dependent on helicopters for the last stage of their supply chain, and for lifting out hundreds of men debilitated by the conditions. The machine of choice was the Aérospatiale Lama, along with an Indian-manufactured version called the Cheetah. For almost twenty years, each side attempted to leapfrog the other, looking for gun emplacements that could shell but not be shelled in return. One solution: the high-altitude helicopter raid. In April 1989 a Lama helicopter carried a squad of Pakistani troops one at a time and dropped them onto a saddle-shaped ridge at Chumik Pass, altitude 22,100 feet, allowing them to sneak up on an Indian post. The Siachen War ended in 2003 with a cease-fire.

Back in the western United States, hundreds of helicopters were mobilizing for a charge into the mountains. The reason was an economic boom of gold-rush proportions.

The first rumblings came in 1974, when a drill rig in northwest Utah confirmed the presence of major oil and gas deposits along the Rocky Mountain Overthrust Belt. The Overthrust Belt follows the Rockies from southern Canada into northern Mexico. Though geologists suspected that the quantity of petroleum trapped there was vast—if developed it would raise America's oil reserves by half—comparatively little drilling among the mountains happened in the following years. Conventional oil-exploration techniques relied on thousands of pounds of acoustic gear; and the trucks that hauled it could not get into remote areas.

This posed a dilemma to wildcat drillers. The hard-rock depths of the Overthrust were so twisted and folded by ancient mountain-building that geologists could not pinpoint pockets of oil and gas with sufficient accuracy to justify the enormous expense of hauling drill rigs into mountains. Then in 1979 the revolution in Iran raised oil prices to historic highs. By this time new computing power and lightweight oil-hunting gear made it possible for geologists to peer thousands of feet into the depths. Further, helicopters were available in profusion to haul the oil-hunters.

While the great bulk of the helicopter activity occurred out of the public eye and ear, by 1980 the commotion in places such as the Star Valley of western Wyoming was enough to attract attention. If late-arriving customers of the Corral Motel in Afton failed to notice the half-dozen helicopters parked behind the log cabins each night, they certainly heard about it the next morning. "The owner liked having us there," recalls helicopter mechanic Randall Sowa, "because all the helicopters would fire up and leave at the same time, seven A.M. each morning, and it woke up everybody else in the motel. That way the cleaners got an early start." This gaggle of helicopters and others converged on Poker Flat Meadow south of town, an improvised landing zone. Each day, pilots shuttled exploration crews, their gear, and explosives into the mountains. Activity was constant throughout the daylight hours. Mechanics stood by fuel trucks equipped with tool chests and spare parts. Says Ken Johnson, a pilot for Continental Helicopters, twenty or even thirty helicopters at a time used "Poker International Airport."

By 1981 the demographics of helicopters working the Over-thrust Belt included egg-shaped Hughes 500s, powerful French-built Lamas, German Bölkows, twin-engine Bell 212s and LongRangers, along with aging Vietnam-era Huey helicopters that been sold off by Air National Guard units and police departments.

Working whenever the weather legally allowed helicopters to fly and sometimes when it didn't, these "seismic" crews flew into the western Shangri-La to lay out long stretches of electrical gear and plug them into data recorders. Then the detonation of small explosive charges nearby generated shock waves like tiny earthquakes. Acoustic sensors caught the waves' reflection. With the help of computer processing, deep underground formations stood revealed at last, like submarines betrayed by sonar.

This description makes the work sound easier than it was. A string of heavy cable and acoustic sensors stretched as much as six miles. After being dropped off on a slope or peak, crews had to manhandle the ninety-pound spools of cable across canyons, up cliffs, into thorny brush that tore clothes to shreds, and down sixty-degree slopes of slippery rock. Back in town, nothing about the work was glamorous either. Workers lived in tents, pickups, and dog-eared motels. They crammed into ranch bunkhouses. One veteran helicopter operator in Casa Grande, Arizona, repaired its Lamas in the back of an old gas station. Despite the blue-collar ambience, the bill for such work was strictly for companies on a champagne budget. A typical helicopter cost $1,000 to $3,000 for every hour the engine was running. Setting out a single string could cost an oil company $1 million.

Still, in the mountains nothing but a well-piloted helicopter could set the workers (called juggies) and their bulky gear into position. While juggies and surveyors rode inside the cabins, most of the hardware and all the dynamite dangled outside, at the end of steel "long-line" cables. Many crashes during the seismic boom originated with pilot mistakes while long-lining. Summer temperatures conspired with high altitude, heavy payloads, and mountain turbulence to make many flights loom as a meet-your-Maker experience, until pilots learned to cope.

"I'll never forget my first tour," says Ken Johnson, who began flying a Bell JetRanger for Continental Helicopters in 1980. "After the first twenty days on seismic it took me a couple of days to settle down, then I started tensing up three or four days ahead of when I went back on duty."

"Lots of ex-military pilots found out that the mountain training and flying they'd done while on duty did not help them much when flying from sunup to sundown in a single-engine machine with absolutely no crew to help them and never seeing the same landing spot twice except when returning to their fuel truck," comments Pete

Gillies. Now a chief pilot and veteran instructor in Rialto, California,*
Gillies worked as a supervising pilot for Western Helicopters during
the oil boom. "The learning curve was very steep."

Ken Johnson came to the work with 1,400 hours of training and
experience as an army pilot, post–Vietnam era. That sounds like a generous amount of blade time, and it included mountain-rescue work in
Washington state, but he found that it was no more than a warm-up.
"What the army teaches about mountain flying is really different than
what you need in the civilian world. With the army there are so many
safeties built in. You're never at maximum gross weight—you never
really work the aircraft. As a civilian it was like learning to fly all over
again, because the bosses and the line guys knew the aircraft better
than the pilots did," Johnson says, "so they asked a lot. I wondered if
this work was for me, this living on the edge all the time."

The helicopter wreckage piled up. In Star Valley, a crash per
week was the average during the peak period, says mechanic Randall
Sowa. But among the risks there were bonuses: Just the simple act of
riding to work in the morning, watching the sunlight burn through
the mists and reach the deep ravines, was exhilarating. When crossing
a treeless plateau or desert valley, veteran pilots flew at 100 miles per
hour just a dozen feet off the ground—to avoid the extra time and
fuel that gaining altitude would require.

By 1984, oil prices had sagged and the seismic surveys were mostly
complete. Though the Overthrust heli-boom was over, piloting jobs
were opening up in another rapidly expanding field: emergency
medical transport. Use of helicopters for transporting injured and ill
civilians began well before the Vietnam War even started,[2] but the use
accelerated greatly as pilots and helicopters became available.
Maryland started a state-supported system of helicopter transport for
all citizens in 1969, operated by the state police and the University of
Maryland, using four Bell JetRangers.** No helicopter fee showed
up on hospital bills in Maryland, because all citizens shared in the

* One of Gillies's specialties is training pilots to handle fully realistic engine-out emergencies so they
can make a safe landing under the most difficult circumstances.

** Some of the Maryland State Police's first generation of flying ambulances later appeared in the
Overthrust Belt, after being surplused out.

cost by paying a surcharge on their motor vehicle registration. Such programs grew rapidly as veteran pilots entered the job market and as the army sought to hand off military-surplus helicopters to state and local government agencies for as little as $100.

Medical helicopters had no trouble finding customers at the garden-variety auto crash site, but they proved particularly useful in saving victims of severe trauma far from hospitals or roads. One case was the wreck of the Amtrak *Montrealer,* a train carrying 278 passengers, early on the morning of July 7, 1984. The location was a hilly area ten miles from Burlington, Vermont. The cause was a storm that had filled a string of upstream beaver dams so they failed all at once, releasing a flash flood against a saturated railroad embankment. Quaking from the passage of two locomotives and two passenger cars at fifty-nine miles per hour, the dirt embankment collapsed, leaving the track without a means of support. One sleeping car plunged into a ravine and was crushed by two other coaches that tumbled in after it. For hours the wreck was reachable only by footpath, pending work by bulldozers to punch a road through the woods. Four helicopters moved fifty of the most severely injured passengers and crew members to nearby hospitals.

Four years after the wreck of the *Montrealer,* 150 helicopter-ambulance programs were running nationwide. Most relied on single-engine helicopters and on Vietnam-era pilots. While the cost was high, averaging $2,000 per trip in California, hospitals saw it as a way to keep the occupancy rate up. It also carried a banner for the hospital. One set of statistics made the point: Nationwide, the mortality rate for the first 1,000 helicopter-lifted patients was cut in half.

But statistics were starting to show something else as well. A rash of fifty-six accidents led to an investigation by the National Transportation Safety Board in 1988. Air-ambulance pilots were being overwhelmed with difficult tasks, often while fatigued: handling up to three radio frequencies and navigating in the dark to an unfamiliar location and landing without assistance, while dealing with the bad weather that often accompanies car crashes and other emergencies. One change was to train aeromedical pilots to wait for police and firefighters to secure the landing zone before the mission launched. Pilots began refusing to fly if visibility was poor. Dispatchers tried to limit the amount of patient information they relayed to the pilot to avoid the White Knight syndrome. News of an injured child, in particular, was thought to tempt some pilots into taking unwise chances. Those pilots, says Robert Steinbrunn, a Minneapolis medical helicopter pilot since 1985, "lost sight of the fact that they were not in a combat situation and we could always send a ground ambulance."

Such risky behavior included flying into bad weather without the necessary navigation equipment and trying to land among trees and wires at night. For his part, Steinbrunn never attempts a night landing by a car wreck until the area has been secured and marked by police or firefighters. The medical helicopter, he says, is for transporting patients and not rescuing them. "It's similar to a chainsaw—highly useful if used correctly, but very dangerous if not."

By this time New York firefighters were having similar doubts when it came to using helicopters on rooftops during skyscraper fires. In his 1977 book, *High Rise/Fire and Life Safety,* FDNY Chief John T. O'Hagan made a case against relying on helicopters to rescue rooftop refugees at high-rise fires. In nearly all cases, O'Hagan said, occupants should either stay in place or (if instructed) take the stairs to street level. His point of view was colored by his interpretation of events during three such emergencies in South America. His book strongly influenced the Fire Department of New York in the following years.[3] But the police department remained confident in the usefulness of helicopters. The dispute became public in 1993, after pilots of the New York Police Department pulled dozens of citizens off the World Trade Center roof following a terror bombing there. Fire officials criticized the helicopter use as dangerous and unnecessary. Some of the tension arose out of the fact that in New York, the police department operated helicopters and the fire department did not; and, further, the police department had an unusually broad role in rescue work that did not sit well with firefighters.* This simmering

---

* By contrast, other big-city fire departments have access to their own helicopters or share the resource with police without noticeable rancor. The peculiar situation in New York dates to the 1920s, when a series of stunt-flying accidents caused the NYPD to muster an aerial unit to nab reckless pilots. The Aviation Bureau also pursued liquor smugglers and carried out water rescues. The bureau meshed well with another NYPD initiative of the era, which was to conduct heavy-rescue work at building emergencies other than fires. The result was police dominion over emergency helicopter work in New York City.

dispute about the role of helicopters in skyscraper emergencies went public once more after the terror attacks of September 11, 2001.

Rescue helicopters have participated at more than two dozen high-rise rescues since 1963. Each of these rescues was unique, and most proceeded under difficult circumstances. Not all went as intended, but helicopters did bring 1,200 people from their perches.

Two of the earliest and most spectacular fires happened in downtown São Paulo, Brazil, in 1972 and 1974. The first was at the 330-foot-high Andraus Building, a concrete-framed structure that held offices and a department store. The fire began in a pile of combustibles stored near the fourth floor and went up via the exterior walls, touching off furnishings with heat radiated through the windows as it climbed. Crowds gathered on the street, blocking fire apparatus, to watch. At its peak the wall of flame was 350 feet high, burning with the fury of an oil-well fire. Three hundred occupants climbed the stairs to the rooftop heliport (the first to be built in São Paulo), after which someone in that group locked the steel door.* But there were 200 more occupants still in the stairwell, and they jammed against the obstruction. Others below pushed desperately upward.

A helicopter with firefighters approached, then banked away when the crowd charged it. But the helicopter came back and hovered out of reach to drop off firefighters, who quickly took control of the crowd. This allowed helicopters to land safely. The helicopters' biggest contribution toward survival that day was not getting people off the roof, but getting firefighters to the roof so they could break open the stairwell door. This certainly prevented many dozens of people on the stairs from being crushed or asphyxiated by the upward pressure. As it was, some in the stairwell had broken bones or were unconscious from smoke inhalation when help arrived. In the end sixteen people died at Andraus, but the toll would have been much higher had the rooftop doorway remained blocked.

São Paulo suffered another skyscraper fire two years later, at the Joelma Building, and helicopters also fluttered to that scene. But the rooftop area available to helicopters was so small, and the fire and smoke so intense, that ninety people died before the aircraft could

---

* Most likely out of fear that more people coming up the stairway would overwhelm the small space available, and make helicopter rescue impractical.

move in. Some survivors told investigators afterward that they had remembered how helicopters had lifted people off the Andraus roof two years earlier and that memory persuaded them to go up the stairs instead of down, because they assumed a similar rescue would be possible. This led to one of O'Hagan's principal concerns: hope of helicopter rescue was more likely to mislead occupants than help them.

High-rise emergencies continued after O'Hagan published his book, each offering new rescue lessons. A 1980 fire at the twenty-six-story MGM Grand Hotel in Las Vegas prompted the biggest gathering of helicopters at any high-rise emergency, before or since. Police helicopter pilots were the first to see the smoke, and sent out a radio alert. Thirty public and private helicopters mobilized to move 300 people from the roof. Coincidentally, air force rescue helicopters and crews were close at hand, having staged at nearby Nellis Air Force Base for an exercise. Fearing that the roof deck was about to catch fire, the pilots organized their ships into a racetrack pattern: approaching from the east side, picking up hotel guests on the roof, departing on the west, then dropping people off in a parking lot. Some guests who waved for help were on balconies well below the roof level, which might seem to be out of reach for helicopters, but rescues proceeded even at that difficult location after a flight engineer lowered himself on a "jungle penetrator" seat at the end of a cable and tossed a rope to those on the balconies. The hotel guests then pulled the engineer to their balcony and, one by one, joined the engineer on the rescue seat. One volunteer pilot was Mel Larson, then a vice president at Circus Circus Casino and owner of a helicopter charter service.

"It was not orderly," Larson says of his first view of the roof, where people were trying to grab hold of the landing skids as helicopters departed. "It was panicky. We had to have police to get control." At one point a police officer drew his gun to make his point. Eighty-four people died at the MGM Grand, but no fatalities happened on the roof.*

One solution to unruly crowds and smoke obscuring the roof, it

---

* For years afterward, and at Larson's suggestion, Circus Circus maintained a cache of rescue gear on its roof in case a high-rise fire required helicopter help.

seemed to fire chiefs at the time, was to keep the rescue helicopter high above, away from fires, where pilots might lose their orientation, sustain damage from heat,[4] or suffer loss of power.[5] In 1976 the chiefs of two dozen fire departments wrote to McDonnell Douglas Corporation and described the problem. The aerospace contractor came back with plans for the helicopter-based Suspended Maneuvering System and went on to build a prototype. The SMS was a sort of supercharged, sixteen-passenger rescue basket designed to hang from a large helicopter at the end of a 500-foot-long cable. The SMS had an engine and thruster nozzles so the operator could move the basket from side to side at the end of that tether, allowing it some freedom of movement when alongside a tall structure or above a flooding river.

The SMS was working by 1978, but despite their early enthusiasm, fire departments dropped the idea. Only one, the Los Angeles Fire Department, actually fielded the SMS, and that deployment lasted less than one year. The chief obstacle was not the rescue device

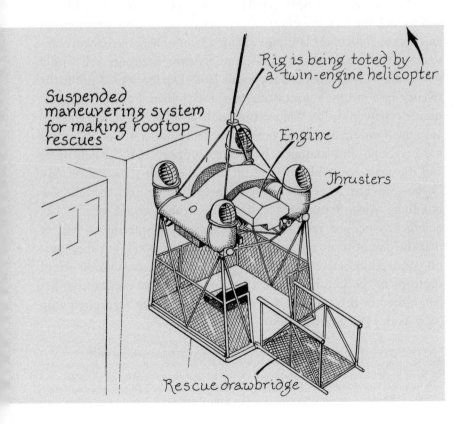

Rig is being toted by a twin-engine helicopter

Suspended maneuvering system for making rooftop rescues

Engine

Thrusters

Rescue drawbridge

itself, which worked well in tests, but the logistical demands it placed on a fire department compared to the rarity of its use. Any department planning on the SMS needed to have a big twin-engine helicopter on call, comparable to the Sikorsky H-60 Black Hawk, because of the weight to be carried and the need for extremely high reliability whenever people are being hauled around at the end of a cable. "I'm sure they would like to have something like this," a spokesman for the Fire Department of New York told *Popular Mechanics* in 1979, "but the dough at the moment is bad news." Also weighing against deployment of the SMS was the scarcity of high-rise fires. The frequency plummeted after big cities began requiring sprinkler systems in new skyscrapers, a move spurred by the disastrous conflagrations of the 1970s. Sprinklers and other fire-minded measures worked so well in catching fires early that if a fire chief had authorized an SMS unit in 1980 he could have worked another two decades and retired without seeing the SMS squad go into action, even once, at a high-rise fire.*

But without an SMS handy at short notice, fire departments had to face the fact that most skyscraper rooftops were unsuited for emergency helicopter landings. They sprouted heat exchangers, antennas of all kinds, water tanks, and elevator housings. All would pose dangers to helicopters trying to land. During a 1987 fire at the twenty-two-story Dupont Plaza hotel in San Juan, Puerto Rico, volunteer rescue pilot Pat Walter could only approach evacuees trapped atop the building by edging one skid onto a parapet. (Pilots call this a toe-in or one-skid landing. The ground supports half the helicopter and the main rotor blades hold up the other half.) Two cities, San Diego and Los Angeles, ordered emergency helipads on all new buildings with floors higher than fire trucks could reach. The emergency-helipad requirement led some helicoptrians to believe that the day of rooftop commuting might be at hand, but there was no connection between emergency helipads and commuter use of rooftops in the United States. One reason rooftop commuting has languished in this country is a chain of events that transpired atop New York's Pan Am Building.

---

* A twin-engine helicopter equipped with the Suspended Maneuvering System, if available on very short notice, would have offered a chance for some trapped office workers to escape the upper floors of the World Trade Center towers on September 11, 2001. The SMS was designed specifically to cope with such a situation. Victims would have edged through broken windows on the upwind side of the building, and stepped onto the SMS's short catwalk and into the cage.

One-skid, or
toe-in landing

In 1926 Thomas Edison gave an interview about aviation that touched optimistically on the subject of helicopters and cities. "It does not require any very vivid imagination to help us realize that when the helicopter comes into being, roofs of large buildings in our cities immediately will become very valuable parts of such structures," Edison said. "Certain new varieties of disaster will then develop, but is useless to foretell these now. We shall know all about them when they come. But they will not keep us out of the air transport machines."

Thirty years later, businessman Horace Brock had doubts about the skyscraper-roof plan. Brock was an executive with New York Airways, a small commuting line that had started scheduling helicopter flights from waterfront helipads on Manhattan Island in July 1953, emulating European programs that dated to 1950.* In 1955 Brock told an aeronautical society that while the idea of a rooftop commuting heliport sounded appealing, it might not work out. The higher the heliport, the more it would protrude into a cloud deck, and the reduced visibility would restrict operations. Elevators would slow the movement of people and baggage, and storage of fuel would

---

* European Airways initiated the world's first successful helicopter airline. It operated in the United Kingdom.

raise fire concerns. Even so, New York had some experience with earlier rooftop heliports* and a grand vision of helicopters on skyscrapers began to take shape soon after Brock's cautionary words.

The space that eventually became the Pan Am Building started out as a concept for an eighty-story tower that would take the place of Grand Central Terminal. In 1958 it was condensed to stand at the back of the terminal rather than replace it. When completed in 1963 at fifty-eight floors, the building blocked the view down Park Avenue and most architectural critics didn't like its plain, beveled-box design. Still, at 2.4 million square feet, it was for a time the world's biggest office building.

It would have another distinction. In September 1960 the president of New York Airways saw a notice about the building plans and wrote Juan T. Trippe, chairman of Pan American World Airways, proposing a heliport and offering to lease it: A heliport, he said, could "be of inestimable value to the community." Pan Am agreed that the opportunity should not be missed and began moving down the paperwork trail. Four years later the plan almost came undone when the commander of the NYPD's aviation unit, William McCarthy, recommended in a confidential report that the city hold off on approving the zoning change for the heliport. McCarthy, who was on the edge of retirement, said the heliport was not a top priority for the city because riverside heliports were available just a mile away to the east and west. Opponents found out and published his report, but the zoning change went ahead anyway. Test flights began in May 1964, followed by promotional trips for VIPs to the New York World's Fair. The heliport opened for business on December 21, 1965.

The historic day began with a lunch at the Waldorf-Astoria's Starlight Roof and a speech by Juan Trippe. Also present: Francis Cardinal Spellman; the president of New York Airways, the operator; the president-elect of the city council; and the president of U.S. Steel. "Urban helicopter travel is here to stay," said Trippe. The helicopter to be used was a civilian version of a military turbine-powered transport that seated two dozen. The airline planned to move over 4,000 people every week. The guests adjourned by limousine to the

---

* A Port Authority bus terminal on Eighth Avenue was designed with helicopter landings in mind in 1950 but it was not used for commuting. The Port Authority did use a rooftop heliport atop its building at 111 Eighth Avenue, for employee travel.

Pan Am Building for the cutting of a red ribbon held by two helicopter hostesses. Vice President Hubert Humphrey telephoned to turn on the heliport lights by "remote control" and to offer a message. "The establishment of helicopter service between city centers and outlying airports can do much to ease urban traffic congestion," he said. "I hope other major cities will soon follow New York's example." Regular service began that evening.

To air-minded futurists, rooftop travel was overdue and a vindication for dozens of Sunday newspaper supplements and magazine covers that had promoted such opportunities. It had been eighteen years since a helicopter first set foot on Manhattan, at the request of a department store. A few other cities had downtown heliports at this time, but only on the Pan Am roof did the reality approach the original vision.

Using civilian versions of the twin-rotor CH-46 Sea Knight helicopter,[6] New York Airways flew 350,000 people to and from the fifty-eight-floor building over the following three years. Flights were short, connecting customers with nearby airports and the Wall Street Heliport.

Because the Pan Am Metroport seems so long ago and far away, the details are worth elucidating. The process worked like this: Travelers pulled up at the Park Avenue northbound ramp, near the Pan Am street-level ticket counter. Helicopter tickets were seven dollars one way. Relieved of luggage at the counter, passengers rode an elevator to the fifty-seventh floor. On that floor a Copter Club served roast beef, salad, coffee, and dessert for $3.50. Musician Jimi Hendrix once held a press conference at the Copter Club.

Two flights of escalators led to the glass-walled staging area, which was set at half height into the roofline to keep passengers and building well below the level of the main rotors. When one of the blue-and-white helicopters arrived and was ready to receive them, passengers left the staging area up a short set of stairs and walked onto the half-acre concrete helipad. The helicopter delivered passengers directly to the Pan Am terminal at Kennedy Airport. Forty-five minutes after getting out of the cab at Park Avenue, passengers walked aboard their jetliner to Europe.

Except for days of fog and cloud, the convenience factor was high. Still, the big turbine-powered helicopters were very expensive

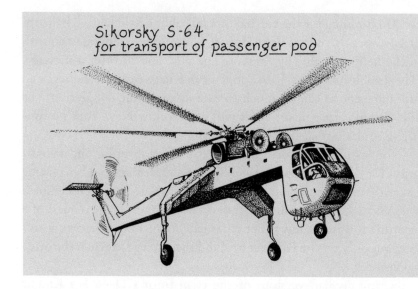

Sikorsky S-64
for transport of passenger pod

to operate and costs outran ticket revenues every year. Despite the predictions of Hubert Humphrey in 1965, the Metroport was still the nation's only commercial, rooftop heliport (others used ground-level helipads). All downtown-to-airport helicopter shuttle operations nationwide had the same problem by 1967: Without subsidies from the federal government or from airlines wanting passengers, they could not continue.[7] Predictions that large, modular helicopters could make the economics work by alternating the transport of "people pods" and cargo containers never came to pass.*

Accordingly, New York Airways halted the Pan Am rooftop flights in February 1968. The glass-walled passenger lounge—set halfway into the rooftop pad as if in a bunker—stood empty and nearly forgotten for the next nine years. A rare and unscheduled visit came in May 1974, when a helicopter hijacker by the name of David Kamaiko directed his hostage-pilot to touch down on the half-acre concrete helipad, pending the delivery of a ransom.** Longtime opponents of the

---

* In Sikorsky's concept, heavy lifters called Sky Cranes would haul bus-like vehicles called Sky Lounges in the morning, move cargo by day, and return to passenger service in the afternoon. Among the cargo that would be moved were sections of prefabricated houses. Sikorsky fielded two versions of the Sky Crane, and the military experimented with two dozen detachable pods for military and hospital equipment, but Sky Crane–carried pods never moved troops into battle or commuters to work. The idea remained alive through the sixties.

** Kamaiko's demand was that a bikini-clad woman be used to bring $2 million in cash to the Pan Am roof. Police officers arrested Kamaiko instead.

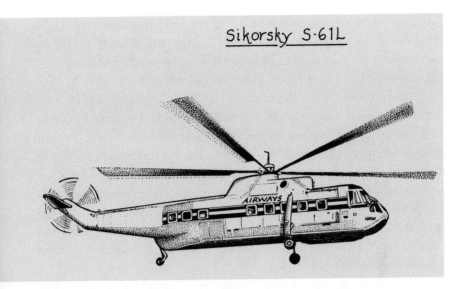

Sikorsky S-61L

Metroport may have regarded this news with a certain grim humor, but Kamaiko was not to be the last passenger on the Pan Am roof.

New business plan in hand, New York Airways resumed flights to the Pan Am Building in February 1977. It had a new helicopter as well: the Sikorsky S-61L, a civilian version of the turbine-powered, single-main-rotor CH-3 Sea King helicopter widely used by the armed forces and the president.

On the afternoon of May 16, 1977, one of those helicopters landed at the Pan Am Metroport. Business was finally looking up: ticket sales were growing 30 percent per month and the operation was projected to clear a profit that very month, for the first time ever. The NYA fleet made four rooftop landings every hour, twelve hours per day. Flight 972 had just come from Kennedy and would soon be returning there with twenty-one passengers. Following standard practice, the pilot left the twin turbines running at low power and the rotors spinning, since the unloading and loading of passengers and baggage into the fuselage took only a few minutes. Then the helicopter would lift off.

Two minutes after touchdown, a fatigued portion of a metal fitting cracked on the right-side landing gear. The gear collapsed and the helicopter rolled gently to the right, instantly killing those caught in the rotor blades' path. The blades hit the concrete deck and hurled fragments into the line of passengers approaching from the staging area. Some passengers were far enough away that they could dive for safety, down the stairs to the lounge. Windows shattered in the lounge. All passengers inside the helicopter were unharmed, because the debris radiated away from the aircraft.

Wendy Goodman was at the Yale Club's library on Forty-fifth Street when she heard a pattering noise like hail; she looked outside and saw a shower of glass fragments. It was from the waiting room atop the Pan Am Building and it continued for three minutes. Blade fragments traveled as far as four blocks north from the building. One section of a blade sailed off the roof, spinning like a boomerang, and looped to crash back into an office on the thirty-sixth floor of the Pan Am Building. A six-foot-long section of blade killed pedestrian Anne Barnecott where she stood—at the corner of Madison Avenue and Forty-third Street.

"There is no conceivable justification for allowing helicopter flights in one of the most densely congested residential and commercial areas of the world," Democratic city councilman Carter Burden told the *New York Times*. Mayor Abraham Beame immediately closed the heliport. "When there's an accident at an airport, they pick up the plane, start the investigation, and open the airport," said New York Airways spokesman Ben Kocivar. "Why can't they do that here?" It was a point worthy of debate, but to *The Nation* magazine, the issue was settled. "They are disastrous vehicles," stated an editorial that month, "their use is limited to the very rich, and they are threats to the lives and health of hundreds of thousands of people around New York City who can never enjoy their supposed 'benefits.'"

New York Airways continued operating out of other heliports after the Pan Am crash, but filed for bankruptcy after a second crash in 1979 that killed three people. The Pan Am Metroport never embarked another commuter after May 1977, and there is no trace of it now on the building's top.* Some of New York Airways' original helicopters are still flying; one is the Sea Knight registered as N6675D, which found a postcommuter job flying cargo for Columbia Helicopters. Having exceeded 50,000 flight hours since its manufacture in 1962—the highest usage in the world—it is the grand old man of working helicopters.

* The Pan Am Metroport operation can be seen briefly in the motion picture *Coogan's Bluff*.

Chapter Thirteen

Chariots of the Gods

When in 1985 the self-awareness expert Maharaji wanted to amend a permit to bump up the number of helicopter landings at his Malibu estate's heliport to thirty-six per year, one regional planner asked at a Los Angeles County meeting, "Why can't he drive like everyone else?" Maharaji's publicity and legal representatives answered, in a diplomatic way, that he wasn't like everyone else. First, the twenty-seven-year-old made many exhausting trips overseas. Once back at the airport, he needed to get to bed promptly. Second, he had received threats on his life, and the helicopter would foil these gunmen. Third, drivers on the Pacific Coast Highway should be grateful rather than resentful because each commuting-copter like Maharaji's meant one less car on the road. The county turned down the request even so, leaving his heliport capped at twelve landings per year.

Similarly, many other helicoptrians who live in resort areas and near big cities have had trouble with neighbors when it comes to helicopter landing rights.

In 2001 a borough government in England turned down a request by Mohamed Fayed, the owner of Harrods, to use a rooftop helipad at his workplace, and refused to budge even after he claimed the heliport would be useful in case one of his employees had a heart attack. Without landing rights in cities and suburbs, business helicopters are of little usefulness on a day-to-day basis.

How did this happen? Helicopter pilots have done many good deeds through the years, and vast numbers of citizens who don't plan to ride in one might someday be grateful for rotary wings if they had a heart attack while camping or got hurt in a bad car crash. Without asking even for gas money, some private owners regularly assist in search and rescues.*

It has something to do with image. An earlier bestseller by Erich von Däniken argued that ancient aliens had handed out secrets of their technology to earthlings, and that evidence of that bounty was simply everywhere. His book's title, *Chariots of the Gods,* could also describe the long-standing association between helicopters and wealth, power, and fame. Celebrity pilots who have flown their own helicopters over the last fifteen years have included Roman Catholic archbishop Roger Mahony of Los Angeles, New York mayor Michael Bloomberg, numerous race car drivers and sports-team magnates, Los Angeles land developer Michael Harrah, Goldman Sachs executive Christian Siva-Jothy, De Beers heir Nicky Oppenheimer, General Motors vice chairman Bob Lutz, and beer czar August Busch III. Many more notables, like Wayne Huizenga, own private helicopters but hire others to fly them. According to the *London Sunday Times,* helicopter ownership in England rose tenfold from 1990 to 2005 for two reasons: highway congestion and wealth.

But English helicopter activity pales with what is under way in São Paulo, Brazil. At least 400 helicopters operate daily across this vast metropolis of 20 million. It has more skyscrapers than New York, and most of the city's 300 elevated heliports are perched atop high-rise buildings. That puts São Paulo in the top three cities worldwide as far as its helicopter population, and top in the number of roof helipads. Helicopters operate with little red tape as long as they stick

* In December 2006, volunteer pilot John Rachor (who was not part of the official effort to find the missing James Kim and his family) used his private helicopter to search the mountains of southern Oregon. He found Kim's wife and children, alive.

to corridors defined by major roadways and if they avoid airliner flight paths.

The first helicopter came to São Paulo in 1948, and now commuting directly to one's business roof is routine for those with the money. Commuters who cannot afford a personal helicopter of their own can rent a seat, through "fractional" packages that provide the customer with a specific number of trips per month. Typically, pilots drop off their passengers downtown, then continue on their way, which leaves the helipads free for other arriving machines to use. Residents in the wealthy suburbs to the west and southwest are the heaviest users.[1]

No antihelicopter movement has taken root in São Paulo that is comparable to those that have sprung up across the United States. In New York, opposition appeared at the first city hearings on the Pan Am heliport zoning change held in 1963. Hundreds of Midtown residents and building owners mailed letters and sat through meetings. They called themselves the Emergency Committee Opposing the Pan Am Heliport, or ECOPAH. One was actress Katharine Hepburn, a resident of East Forty-ninth Street. Hepburn said at a hearing, "New York is noisy and dangerous enough without this nerve-wracking and fundamentally unnecessary addition of a luxury trade operating in the center of the city to the distress of many and to the benefit of the few." Legendary public-works builder Robert Moses entered the fray in 1967 with a *Times* op-ed piece blasting Braniff Airlines for a newspaper ad that suggested that the Pan Am helicopter route was preferable to car travel on the Long Island Expressway. ECOPAH also warned, somewhat presciently, of helicopters plummeting into Grand Central Station.

The final closure of the Pan Am Metroport in 1977 proved to be only one checkpoint in a longer dispute about helicopter commotion. Of those, noise is at the top of the list. Except during extreme emergencies such as the Katrina response, the noise problem is a drag on helicopter activity of all types, including heliports strictly dedicated to public service.

Eurocopter has announced plans to cut main-rotor noise by six decibels on its larger models by adding computer-driven electric motors that will change blade pitch twenty times per second. The rotorcraft program at the University of Maryland is experimenting with tiny tubes added to the tips of rotor blades. The current industry goal is to cut noise from commercial helicopters to at least seven decibels below what federal rules require.[2]

Noise has been the principal driver behind local opposition to helicopters around the country ever since 1947, when the world's first rooftop heli-commuting service, Skyway, got under way in Boston with a Sikorsky S-51. Each Skyway flight carried three customers to and from Logan Airport. The noise inspired residents and managers of hotels ringing Park Square to write letters to politicians.* Nationwide, there are at least three dozen antinoise groups, including those in Hawaii, New York, Washington, Colorado, Alaska, Florida, New Jersey, California, and Arizona. One flagship group is the Helicopter Noise Coalition in New York City, which has lobbied the city to close two heliports and wants to cut off commercial flights from the rest.

One of the now-defunct heliports was operated between East Sixtieth and East Sixty-first Streets, at the East River. New York converted the heliport from a barge-loading station** in November 1968. At the time, Pan Am World Airways promised to run a scheduled service and also a "taxicab" service from there, guaranteeing a copter within a half hour of the customer's call. The main job of the helicopter shuttle would be to allow New Yorkers to fly directly to airports, large and small, such as the one at Teterboro, New Jersey. This, it was said, would take pressure off LaGuardia Airport. A small number of wealthy residents found that for a fee they could park their helicopters there like cars at a garage, which was (and still is) a rare service inside big cities. Police and medical pilots favored it as well. During a peak in helicopter use during the early 1980s, Manhattan had four heliports operating. Flights to and from the Sixtieth Street Heliport hit a rate of 20,000 per year in 1981.

The Sixtieth Street Heliport was a 300-foot-long strip of pavement at the foot of the Queensborough Bridge. It was just ninety feet wide. Cars using Franklin Roosevelt Drive passed a few dozen feet away. The heliport was low and small, while all the structures around

---

* The heliport was on the roof of the Motor Mart parking garage. Actress Lucille Ball was among the customers before Skyway shut down, after four months of operation.

** The New York Department of Sanitation used the wharf to dump snow in the East River and garbage into barges.

it were near and high, except on the river side. One of those buildings was a high-rise hospital for cats and dogs.

"When you came out, you were facing an unknown wind," actor Harrison Ford says of the heliport, which he used often for flying lessons in 1998 and 1999. "The wind was usually along the East River, out of the south to the north. It usually needed an immediate pedal turn upon lifting up. It was a terrible place for a heliport." Pilots compared it to taking off from inside a shoebox that had only one open side.

In 1968, when the bridge-shadowed heliport opened, Mayor John Lindsay promised Rockefeller University, New York Hospital, and many neighborhood residents that they would not experience any "appreciable increase" in noise. But those residents did, recalls Joy Held of the Helicopter Noise Coalition.

"We've been taken hostage by the industry and we can no longer live and work here in peace," Held says. "Wherever helicopters are flying, community groups spring up in protest. Why is that? Because helicopter noise is extremely disturbing on the ground and local jurisdictions have no authority. The important point is that this is not just about the noise. It's also about fumes, the risk to health, and the risk to national security."

I met Held at her apartment house on the Upper East Side, a solidly constructed brick structure with gleaming wooden floors that is over a hundred years old.[3] An airspace in the center of the building admits light but also lets in the sound of helicopters. I hear one fly over, despite the double-glazed windows. In October 1996, Held took a seat on her roof for three days to count helicopters. She recorded thirty per hour for much of each day.

The Noise Coalition's NOMBY (Not Over My Back Yard) sentiment may seem like a modern form of unhappiness, but the mother of antinoise activism lived long ago. She was Julia Barnett Rice, of Manhattan's Eighty-ninth Street and Riverside Drive. Julia Rice founded the Society for the Suppression of Unnecessary Noise in 1906 and was the wife of inventor and investor Isaac Rice.[4] Her chalet-style mansion called Villa Julia was on the opposite side of the island from Joy Held's apartment, overlooking the Hudson River. Her first and most famous protest was against tugboat officers who worked the waterfront of New York and New Jersey late at

night and who had the habit of summoning their crews from the sa-
loons by laying on the steam whistles. At the height of the struggle,
each night tugboats gathered in the Hudson River within range of
her house, whereupon the crews blew whistles and blasted spotlights
into her windows. She won passage of a federal law authorizing the
inspector of steamboats to enforce a whistle ban. Without such action, Julia Rice liked to say, the twentieth century would be the Age
of Noise.

That century has yielded to another, but to Joy Held, the Age of
Noisemakers hangs on. "The Eastern Region Helicopter Council developed a fly-neighborly program, but we feel if we can hear them
they're not being good neighbors," Held says. Her group wants helicopters other than those on public-safety missions to be banned from
the remaining heliports. If private helicopters wanted to go past New
York at existing altitudes, they'd have to detour five miles to sea. Had
she ever ridden on one of the banes of her existence? Yes, once, Held
recalls: it was a helicopter airline operating in the San Francisco–
Oakland Bay area.*

For all her dislike of noise and risks from commercial and private helicopters, Joy Held is not making war with the woman who
once managed the Sixtieth Street Heliport. Her name is Patricia
Wagner, and now she runs another heliport farther down the riverfront, where I find her on a wet and blustery Friday evening. Her
heliport is at the end of East Thirty-fourth Street, positioned between an elevated highway and the East River. The five helicopter
parking spots lie in a row along the dock, alongside a metal-walled
building for the office and waiting room. It's one of three public heliports that operate on Manhattan Island.

Though world leaders, business magnates, and giga-celebrities
parade through its doors, the furnishings at East Thirty-fourth do
not aspire to luxury, or not very far in that direction. A set of cast-
iron-framed stools in front of the flight-operations counter looks to
be borrowed from a patio set. Wagner has draped curtains along the
wall, to obscure the rusty view of FDR Drive, an elevated, steel-
framed highway that overshadows the heliport. The heliport build-

---

* San Francisco & Oakland Helicopter Airlines operated a shuttle in the Bay area from 1961 to 1985.
SFO flew to three airports as well as downtown Berkeley and Oakland. The line was notable for
surviving without federal mail-hauling subsidies.

THE GOD MACHINE 219

ing, while tidy, is a metal-walled structure of the mobile type used for temporary offices and classrooms, and not the stuff of young executives' dreams. The building floor vibrates with each big truck. But it's in much better shape than the Wall Street Heliport was in the late 1970s. That one suffered mightily from holes and even a minor collapse, when the end of its rotting pier slumped into the water.

After the Pan Am heliport closed in 1977, all New York heliports have operated on the waterfront; these raise fewer noise complaints than inland ones, since arriving and departing flights can pass over water rather than apartments and hospitals. Its spot on the East River harkens back to one gloriously audacious estimate worked out following World War II on behalf of New York City. A planner presumed that a hundred thousand personal helicopters would be arriving each workday morning in New York and lingering during the business day, in the fashion of commuter cars coming over the bridges or through the tunnels. He figured that the thirty-two-mile length of Manhattan's waterfront, if entirely covered with helipads, would serve to host them.

Access to the heliport at East Thirty-fourth Street is gained via a buzz-through door, but there was no X-ray machine or screening at the time I visited in 2005. The heliport is owned by the City of New York, and leased to a management company that provides services. It first began hosting helicopters in June 1972.

Wagner, dressed in a pink shirt and jeans, has the welcoming style she learned in her years with Braniff, TWA, and Pan Am airlines. She says that the Sixtieth Street Heliport closed for reasons beyond antihelicopter politics. There were safety concerns following residential towers that went up on nearby Roosevelt Island in the East River. There was the proximity to the Queensborough Bridge, and a high wall that interfered with operations.

Wagner attends neighborhood and borough meetings regularly for the Eastern Region Helicopter Council. For New York City politicians, she says, "heliports are a political noose. No one's cheering for us. Whatever you get, you have to fight for." She does not retreat from this platform: helicopters, and not just those flown for public-safety purposes, have a right to operate in and around big cities. When well managed, she says, helicopters are remarkably safe. "For the number of operations annually," she says, "we have the safest record of any other moving vehicle in the world, including boats and planes."

Over the years, she says, helicopter operators have met residents at least halfway by cutting back on night flights and weekend operations, have halted flights that took shortcuts across Manhattan Island at Central Park, and have hiked their cruising altitudes. "Now helicopters cross Manhattan only at fifteen hundred feet or more," Wagner says, "and usually they go around the island. That's our 'fly neighborly' program. To groups I say this: If you want us to say we're going to close, then we're not interested in speaking to you," Wagner told me. "But if you are interested in amelioration, we will talk if we can honorably do it."

Twenty-five years earlier, New York Helicopter operated a commuter service out of the heliport. "There is no scheduled helicopter commuting here now," Wagner says, "but anybody could do it. You could do it, if you wanted to lose a lot of money. I'm just a gas station for helicopters. I don't care where a guy is from or where he's going." Her company makes money by charging landing fees and by selling fuel. Aircraft are welcome to park, for an hourly fee, except for the smallest ones, which can be damaged by rotor downdraft from the largest ones.

The best time for helicopter-watching at East Thirty-fourth would be late on a Thursday afternoon in the summer, when many executives stream out of Manhattan for a long weekend. Common destinations are Morristown, New Jersey; White Plains, New York; and the Long Island communities collectively known as the Hamptons.

But this is October and wind is carrying a cold rain this Friday evening. A lone Sikorsky S-76 appears over the river, lands smartly, and rolls to a parking spot. I gather that it is coming to pick up General Electric executives. Two men arrive in a limo and bustle into the heliport office. One looks like CEO material. The other pauses just long enough to comment that the helicopter trip will save him two hours on the road, and then they depart through a second doorway, onto the helipad. They ascend the foldout stairs of the helicopter. The craft is tastefully unadorned with any corporate name, as is now the fashion with executive transportation.*

* One helicoptrian to wear his name with pride is Donald Trump, who arrives for appearances in a black helicopter bearing his last name in capital letters.

After a lengthy delay for air-traffic clearance, the helicopter spins on its wheels to face the river and lifts off, its anticollision beacons flashing. The rotor noise peaks during takeoff but the noise fades rapidly once the machine is over the water. In seconds I can hear the evening serenade of big-city road traffic.

Leaving the lounge, I notice promotional posters for charter helicopter services, framed and hanging on the wall. From the posters, one could infer that that power-suited men and women use their executive helicopters only during spectacular summer sunsets. The cause of helicoptrians who want to protect the endangered species of big-city heliports has not much been helped much by the love affair between helicopter travel and the rich, famous, and powerful, harkening back to the autogiro's strength among the sportsmen and the smart set. When Tom Cruise wanted to promote his action movie *Mission: Impossible III* in a daylong extravaganza across the city trumpeted as Mission: NYC, Cruise's chartered helicopter used this facility as a launching point to fly to another spot on the waterfront, where he hopped on a high-powered motorcycle.

Any follower of celebrity lifestyles is likely to come across chatter about the latest mega-yachts, the most worthy of which feature a helicopter, a hangar, and a helipad. Microsoft cofounder Paul Allen's "giga-yacht" has pads for two helicopters. In that respect a portion of the early, optimistic pronouncements about personal helicopters has come true, as shown first in Sunday newspaper supplements and popular magazines. The luxury-market term for such customers is the "emancipated wealthy." For the great majority of people, however, helicopters are entirely out of reach. Anyone wanting to earn a private helicopter pilot's license should budget at least $15,000 for lessons, and then plan on $300 per hour renting a small one for personal use.

"I like to ask, what makes a helicopter fly? It's not production of lift, not combustion of hydrocarbons," says medical helicopter pilot Robert Steinbrunn. "It's money."

Since listings for helicopter showrooms do not appear in yellow pages like automobile dealers, people shopping for one are likely to drop in to the annual HeliExpo trade show organized by the Helicopter Association International. Each helicopter maker puts its top-line executive models on display, with a leather-and-teak look perfect for getting out of town on a Thursday afternoon.

Such high-end machines do not roll off an assembly line like cars. Factories such as Agusta, Bell, or Sikorsky leave lime-green metal showing inside the cabin and equip them with only essential instruments for the ferry pilots who pick them up. Accordingly, these are called "green ships" and must pass through a "completion house." One of these is Keystone Helicopters of West Chester, Pennsylvania. Depending on the customer, completion could mean the installation of custom cabinetry, an entertainment center, cabin windows that can turn from clear to opaque, teleconferencing, satellite communications, and an external video camera.

"If a guy gets out of a thirty-five-million-dollar bizjet, he expects equivalent quality, workmanship, and equipment in his helicopter," says Rick Hinkle, vice president for program development at Keystone. The total delivered cost for a new executive helicopter, an Agusta-Bell 139 that seats up to fifteen, can run more than $8 million. The final delivery sequence of such an aircraft takes two weeks, and ends with a red-carpet ceremony.

As Jim DeMarr, a helicopter-services manager for Keystone, advises, the helicopter is not a product for the economically faint of heart, whether being purchased or brought in for maintenance. Rotor blades for the largest late-model Sikorsky, the S-92, cost $250,000 apiece if damaged by mishap, and there are four blades. "But this is not the business to ask about costs," DeMarr says. "If you need to ask the costs, you shouldn't be in it." DeMarr knows his helicopters from the inside out: while in Vietnam one of his duties as a Chinook mechanic was to go into combat zones and make repairs to battle-damaged aircraft so they could get back to base; this was necessary because no other helicopter was big enough to haul them back.*

Even for the smaller Sikorsky S-76, replacing all the blades with new ones would cost a half-million dollars. It's expensive enough that if Keystone comes across a crack in the outer skin that is within limits, the blade can stay in service. He showed me a crack in one blade's leading edge: it was short and perfectly straight, barely visible to the eye. He taps around it with a coin: near the hairline crack, it makes a dull sound. "The book says it's okay if there is no de-bond at

---

* By contrast, a Chinook could haul back a damaged Huey. Crews removed the rotor blades and attached a cable through a ring atop its rotor hub.

the end of the crack. We put clear blade tape on it to keep the moisture out. Blades are so expensive and this is within limits."

Similarly, helicopter fanciers work hard to contain costs, and not all of them are millionaires. At a helicopter air show I met owner Keith Grant, who bought a renovated Vietnam-era turbine helicopter at substantial savings over buying a new model.[5] Aeronautical engineer Frank Robinson has done as much as anyone to cut the cost of small, personal helicopters. If the yardstick is sheer numbers of whirlybirds out the door each year, Robinson Helicopter Company is the largest manufacturer in the world, with less than a thousand rotorcraft produced in 2005. Robinson, born in 1930, went to college as a mechanical engineer, working on oceangoing freighters to pay his tuition. After graduate school he worked for most of the helicopter manufacturers and developed a professional specialty in tail rotors. At each stop, Robinson pressed his employer to consider making a simple, reliable helicopter with room for a pilot and one passenger. None of the established firms acted on his offer, so he went into business on his own in 1973, working out of his home in California. Customers bought his aircraft for training new pilots, business commuting, cattle herding, scouting for fishing boats, and, most recently, for use in police and reporting work because the price of his gasoline-powered helicopters is low compared to turbine-driven helicopters.

Even so, the no-frills version of Robinson's four-seat helicopter, the R44, costs $364,000. Some novices find it challenging to master,* and in no way does it approach the "housewife helicopter" promised by Charles Kaman and many other helicoptrians prior to 1950.

Americans' deflated dream may have contributed to unhappiness during noise hearings. But the ire raised by all those subjects—wealth differences, irritating noises, and imposed risks—are tame when compared to the emotions that the subject of helicopters trigger among a small but vocal group of citizens (and at least one former congressperson).** It regards dark and spooky machines that fly by night.

---

* Any purchaser who wants to buy low-cost insurance arranged by the manufacturer must go through a rigorous four-day training program at the factory.

** Representative Helen Chenoweth, now deceased.

Chapter Fourteen

Something to Watch Over Me

To set the record straight, there really *are* black helicopters, they are run by the federal government, and their crews really do prefer to work at night. I saw rows of raven-colored Black Hawks and Little Birds when visiting the 160th Special Operations Aviation Regiment headquarters at Fort Campbell, Kentucky. Their MH-60, AH-6, and MH-6 helicopters were painted dull black and the numbers and letters on their sides were a barely readable dark gray. And I joined the legion of darkness myself, for a few hours one January night in 2006. I donned helmet and night-vision goggles and rode in a black Black Hawk as it banked serenely* along the Ohio River. I marveled as pilot Brian Lynch demonstrated a shuddering "tactical entry" stop over a small airfield near Tell City, Indiana.

But what about the spine-tingling part—that is, some citizens' bedrock-solid conviction that black helicopters are oppressing farmers and ranchers from

---

* I was also wearing noise-canceling headphones. Noise-wise, the Black Hawk is not a stealthy helicopter.

Idaho to Arizona nightly? And, perhaps more productively, how have real helicopters performed when it comes to ordinary citizens, who rarely get to ride in one? Are today's choppers the boon or bane of democracy?

By 1965 it was common knowledge to watchers of television news and readers of newspapers that helicopters were striving to be the winged nemesis of Vietnamese insurgents. But it was not a common idea among Americans that helicopters might be turned against them also. That slow-growing fear may have taken hold after the Watts-neighborhood riots of August 1965 in South Central Los Angeles. During that crisis, helicopters of the Los Angeles Police Department and KTLA-TV hovered over scenes of violence and looting. During subsequent demonstrations marking the anniversary of the riots, the LAPD launched its helicopters to watch for trouble and, residents suspected, to photograph demonstrators.

Within one year after the Watts riots, cities began expressing interest in using helicopters to stamp out urban violence. Lakewood, California, acted first by setting up Project Sky Knight in coordination with the Los Angeles County Sheriff's Department. Authorities announced that Sky Knight reduced street-crime rates by 8 percent, even as crime rates elsewhere in the Los Angeles area rose by 9 percent. In time, over four hundred cities would follow suit.

Like ripples spreading across a pond, concerns over domestic use of helicopters moved from the African American community to a broader audience because of an extended crisis in May 1969 at the Berkeley campus of the University of California. It began at a weedy plot of land about three acres in size, two blocks from the campus perimeter.

Two years earlier the acreage had been a run-down block of brown-shingled apartment buildings populated by students and hippies. The neighborhood held a funky charm for many young people, but state legislator Don Mulford called it a "human cesspool" and pressured the university to buy and flatten all the buildings. The university began buying buildings on the block that year, evicting the tenants, and tearing down the old apartments. By spring 1969 the block was mostly empty of buildings, and it was serving as a muddy parking lot spotted with brush. Concluding that the university's promise of new construction had been a ruse to drive out unwanted residents, activists made the issue a rallying point for their politics: the university had lied and the land would be better used as a People's Park, planted and furnished without help or interference from the Establishment. Possibly hippies could in time use it for parties and concerts. Certainly radicals could use the controversy as an organizing tool.

The *Berkeley Barb,* an underground newspaper, published an article by Stewart Albert that argued for the park plan. The article called for volunteers to turn out with shovels on Sunday, April 20. It would be a proletarian project, without supervisors or a blueprint. The university announced a counterplan for a soccer field, hoping to head off the movement, but was ignored. Hundreds of people showed up on the appointed day, most pursuing pet projects. Neighbors protested the noise and the crowds, but the work went forward. Volunteers slept at the park each night to guard against a play by the police. After three weeks of escalating tensions, early on the morning of May 15 the university and law officers moved in. This was expected to resolve the issue and allow the construction of a fence before an upcoming university regents' meeting.

Officers and national guardsmen evicted dozens of sleeping hippies and radicals while a helicopter monitored from above. Despite a ban on all rallies and assemblies signed by Governor Ronald Reagan, tensions spiraled. Handbills passed out on the street called for a demonstration at the university's Sproul Hall at noon the same day, which at least two thousand people attended. Given the ferment of the time, it wasn't hard to raise a large crowd in Berkeley; some sorority women liked to turn out for riots because it was a way to meet young men.

Halfway through the roster of speakers, student Dan Siegel called for residents to take back the park. Siegel may have been making only a rhetorical point, but the defiant words "Take the park!" spread like a detonation wave through the crowd. The event broke up with a roar as thousands headed for People's Park via Telegraph Avenue. There they met Alameda County Sheriff Department deputies, campus police, and guardsmen. Scuffling broke out and then low-intensity urban warfare. Demonstrators burned civilian autos and patrol cars. Concrete blocks and bricks tumbled down from rooftops. Deputy sheriffs used their guns and tear-gas canisters. National guardsmen followed demonstrators to other "People's

Park Annexes," and, following orders, plucked each freshly planted flower.

On May 20 faculty members and students came up with the incendiary idea of staging a "funeral march" to call attention to the fact that one student had died from gunshot wounds sustained in the first riot. They gathered at the chancellor's residence to shout denunciations. As guardsmen in gas masks cordoned off gates to the campus, a Choctaw helicopter swooped in to dispense streams of the white, powdery tear-gas agent called CS. The effects were felt as far as a half mile away, by patients inside Cowell Hospital. It was 2:00 P.M., when schoolchildren were heading home.

"There's no question that the timing was somewhat heinous," Rhys Thomas recalls. "It caught a lot of people who were not involved in the march." Thomas, now a Los Angeles television producer, was gassed twice as a twelve-year-old in Berkeley. CS agent*—the same chemical then being used by the army to clear Vietcong from bunkers in the Mekong Delta—was "extremely unpleasant. It starts with burning eyes, then goes into the nose and throat. It can even cause loss of bladder and bowel control. It really hit hard," he says.

The university's faculty called a boycott the next day, canceling hundreds of classes. The official campus newspaper declared the university closed on the grounds that it was "unsafe for human beings." Questioned about the dispersal of CS from a helicopter, the commander of California's National Guard said it was perfectly logical.

Local anger cooled with time but erupted once more when Berkeley police proposed to buy two helicopters. The city council presided over a series of stormy meetings, interrupted by stink bombs, shouting matches, and accusations that police plainclothesmen were recording the identity of protesters. The ACLU argued that observations from helicopters would violate constitutional rights of privacy. Activists claimed that "whirly-pigs" were going to use their "pork-choppers" to shoot residents and drop more CS. The next decade, which brought Watergate-era worries about government malfeasance, only reinforced a feeling that the government might have used, and was still using, more tools than it was revealing to the public.

---

* Often called tear gas, CS is actually a fine, clinging powder.

Though the full set of social currents that fueled the phenomenon can never be known, the concept of blacked-out, night-flying helicopters with strange powers first circulated in print in 1973.

That's according to Professor Bill Ellis of Penn State University, an expert in modern folklore and conspiracy theories. Ellis believes that the foundation for the black-helicopter cult was in place by 1965, when it was common for pro-UFO books to cite a claim from a Kansas rancher published in an 1897 newspaper. The rancher had said that a mystery dirigible had lassoed one of his calves, dragged it up and away, and left it mutilated on the plain.* In the pre-Internet era, the books' message lay mostly dormant until 1973, a time of unusually high beef prices brought on by cattle shortages. Then, as news reports emerged about cattle rustling in the Midwest—some of which implicated henchmen in helicopters—it was an easy leap of logic back to the nineteenth century, to mystery airships of 1897 showing lights and ladders. Now the buzz took hold and gave the aerial rustlers special powers. Those weren't ordinary helicopters: they were silent, they disappeared when chased by conventional aircraft, they spurned hoist cables and grabbed cattle by teleporting them instead. They had hypnotizing rays; they could kill grass; they could stop car engines. One chased teenage girls in eastern Colorado.

Meanwhile, genuine incidents featuring night-flying helicopters, particularly those carrying out unannounced antiterrorist training missions, kept the base fully energized. One of these happened on the night of February 8, 1999, at Kingsville, Texas, when occupants of the Casa Ricardo retirement home were startled to see four helicopters touch down on the roof of an Exxon office building across the street. Kingsville residents had not been told of any training plans[1] and were greatly startled by the sounds of exploding grenades and automatic gunfire. In the distance another helicopter hovered over an abandoned police station, its running lights extinguished. The Libertarian Party issued a press release demanding that the army stop "invading" small towns. Conspiracy theorists said Kingsville was but a dry run for the invasion already planned for the Y2K zero hour: December 31, 1999.

Allegations of hushed, night-flying helicopters persist today,

---

* Though the rancher later recanted his report as a hoax, few UFO books mentioned that part.

echoing from Web site to Web site and occasionally popping up at public hearings. Air-ambulance pilot Robert Steinbrunn recalled a neighborhood meeting he attended in Minneapolis to discuss noise-reduction measures at his hospital's heliport:[2] "One guy said, 'What about that Whisper Mode, why don't you use that?' So I said, 'That's only available in Hollywood.' And he said, 'I don't care where you have to go to get it, just get it.' "

Notwithstanding the compelling special effects of *Endangered Species* and *Blue Thunder,* there is no Whisper Mode that can muffle any full-sized helicopter hovering in one's backyard. Part of the blame falls on the noisy and fuel-wasting phenomenon called blade vortex interaction. It happens like this: As the rotor turns, tiny tornados called vortices tear loose from each blade tip. The next blade to arrive tears through those whirlwinds. This makes a window-rattling, low-frequency slap that, with larger helicopters, can be heard for a mile or more.* Blade vortex noise is particularly loud when a helicopter is on final approach.**

Even though blade vortex interaction makes totally silent full-sized helicopters impossible, they can be made quieter than is generally known. Probably the quietest turbine helicopter ever fielded was the limited edition Hughes 500P, dubbed the Quiet One, which the CIA put together for a secret mission to wiretap North Vietnamese telephone lines in 1972. Silencing was achieved by adding an extra main rotor blade (which allowed for a lower rotor speed), by redesigning all the blade tips to reduce blade vortex interaction, by redirecting the engine exhaust, and by redesigning the tail rotor. It also needed an engine muffler and sound blankets on the engine compartment. How quiet was the 500P? Off the shelf, Hughes 500 helicopters could be heard a mile away, but the average person would not hear the Quiet One when it was 500 feet above his neighborhood.[3]

The foregoing seems to add up to a bleak picture: Some people feel threatened by helicopters silent, dark, and deadly, and others are annoyed by ones visible and loud.

---

* Vietcong soldiers could detect Huey rotor-blade noise at a distance of five miles by digging a bowl-like hole in hillsides, which focused sound as a parabolic reflector. These men were called LZ-watchers.

** Most BVI noise from a moving helicopter is cast in the forward direction, as if from a bullhorn, because the advancing blades throw the energy forward and somewhat to the right.

But for years helicopters have been carving a niche that was never predicted for the aircraft, even during the headiest days of helicopter euphoria. That unexpected role is the hovering, omniscient eye—not for the gods, but for the public. "There are no secrets from a helicopter," says Dick Hart, whose company, National Helicopter, put together the first Telecopter for KTLA-TV in 1958.

That public-eye role mostly involves the transmission of televised images, but it can include simple still pictures, such as those taken by Ken and Gabrielle Adelman of Corralitos, California. This couple volunteers their time and their Robinson four-seat helicopter to photograph and rephotograph every mile of California's coastline. Under California law, the beach is owned by the public at large, and modifications such as seawalls must be reviewed rather than built by property owners on the sly. Thousands of Adelman photos, which take months of work to collect, go on their California Coastline Web site for anyone to examine.* The vast collection of new and archived information allows the Coastal Commission and conservationists to catch illegal activity that degrades the coast. Entertainment megalegend Barbra Streisand valued her privacy enough to sue the Adelmans in hopes of forcing them to pull photos of her Malibu estate from the Web site. She lost.

It's all part of nongovernmental surveillance in the public interest, an entirely new form of reporting that appeared only after rotary-winged craft made it possible.[4] Residents of Southern California saw one example in April 1996, when a helicopter was overhead to broadcast the shocking conclusion of a high-speed, seventy-mile pursuit. The copter-cam showed Riverside County sheriff's deputies grabbing two immigrants from a pickup and beating them with clubs, though the two were putting up no resistance to their arrest.** The evidence in such cases is likely to be compelling to a jury because the images are so clear, whether by day or night. Gyro-stabilized telescopic cameras allow newscopters to catch readable license-plate numbers from a distance of two miles.

* At http://www.californiacoastline.org.

** Among other such events captured by newscopter are post-car-chase beatings of suspects by Philadelphia police officers in 2000 and by a Los Angeles police officer in 2005.

Copter-cams have caught outrages committed by others as well, and productively stirred up the citizenry. On April 26, 1992, Los Angeles resident Bobby Green saw television images of five young men pulling a truck driver from his vehicle at an intersection not far from Green's location. Until this bulletin, the news in South Central Los Angeles had been full of an acquittal of police accused in the Rodney King police-brutality trial. It was generating scattered violence. Like Rodney King, Green was African American, and like thousands of others was angry about the jury's acquittal of four white police officers accused of beating King. Regardless of that anger, Green couldn't tolerate what was happening to the white trucker, Reginald Denny. The camera now showed Denny lying on the pavement. A man was beating him in the head with a slab of concrete.

"That is enough," Green said to his family. He left the house, got in his car, and drove to the intersection. Green found a woman trying to lift Denny back into the cab of his truck. He helped her hoist Green in, then drove the semitrailer rig directly to a hospital with the help of shouted directions from two other men who drove alongside in a Honda. Green needed help because he couldn't see where he was going: the windshield was opaque with fractures. The truck reached the hospital as Denny was going into convulsions. Doctors there told Green that Denny would have died from his injuries had even a few more minutes passed.

Green arrived to rescue Denny because three news helicopters were broadcasting live near the intersection. One was flown by Bob Tur. As he narrated the scene for viewers at KCOP, his wife, Marika, shot the video from the open door. After Tur testified at the trial of the four men who beat Denny, prosecutors gave him the concrete slab as a memento.

His participation in the trial was no surprise to those who knew Tur, a man who is far from reluctant to get involved. Tur began working as an ambulance driver and then worked as a sort of *ronin* among television newsgatherers with his wife, Marika Gerrard. Together they were Los Angeles News Service. Following the call of the police and fire scanners, Tur and Gerrard roamed the mean streets each night with an Ikegami camera, taking freelance footage of crime scenes and fires, working only after local-news and network camera crews were snug in bed. He sold the footage each morning to the news outlets.

"We were two skinny white people, with all this gear," he recalls of his nights on the street. "The gang members would come up and ask us if we knew the anchormen, like Connie Chung and Dan Rather. I had been on ambulance work and knew how to maneuver through the tough areas."

At age twenty-one, with $50,000 of videotape earnings but no helicopter license, Tur purchased his first news-copter, a Bell JetRanger, in 1980. Tur paid the fire department's toughest instructor pilots to put him through the new-pilot wringer, then went to work in the air. His trademark product became high-speed police pursuits, capturing them initially by videotape and then by live broadcast, using a microwave link.

Tur's was the first helicopter on the scene on January 3, 1992, when Darren Stroh was evading the California Highway Patrol in a Nissan he had hijacked by killing the driver. The chase lasted for more than four hours. Stroh caused a string of accidents and stole another car, this one a red Cabriolet. At one point Stroh tried to hold off the Highway Patrol by firing a sawed-off shotgun out the rear window. Tur's camera caught the gunfire and pursuit, which reached speeds of 100 miles an hour and crossed over every lane the interstate offered. He relayed it all to KCOP-TV, Channel 13 in Los Angeles, which preempted a rerun of *Matlock* to show everything live. When the station attempted to switch back to *Matlock,* two hundred viewer calls convinced the manager to stay on the pursuit. It ended when Stroh's car ran out of gas south of Los Angeles and coasted to a stop. A patrolman shot him when he delayed in giving up his shotgun. Viewership ratings for KCOP hit such a peak that day that all stations scrambled to ready themselves for the next car chase.

So it was that by June 17, 1994, all the major stations in Los Angeles had helicopters available, along with the cameras and relay links to broadcast live, in flight. Alerted by a phone call from a friend of O. J. Simpson that the former football star was threatening suicide, the police and all the Los Angeles news stations began looking for a white Bronco.

Bob Tur guessed that Simpson was thinking about his wife Nicole's death and therefore would be heading for the cemetery where she was buried. His was the first news-copter to find the Bronco moving at a slow pace on Interstate 5 where it merges with the San Diego Freeway. Thirteen more helicopters joined in.*

Summoned by the coverage, which was viewed live by an estimated 95 million Americans, thousands of Californians lined up along highway bridges to become part of the spectacle, which spanned more than sixty miles. It was more of a mass-participation event than news, and ranks at or near the top of any lists of most-watched television moments. "It's not enough to beat the competition," Tur told *People Weekly* afterward. "You've got to make them miserable."

"What I did was show that you can do police pursuits," says Tur. "Four times we showed people shot; we showed two hundred and eleven car chases. We showed planes crashing on live TV."

Since the O.J. spectacle, the televised pursuit has become a minor industry. It is so institutionalized that a firm called Pursuit Watch sends out an electronic alert to TV watchers whenever a helicopter-car chase is under way in the Los Angeles Basin. These are so frequent that police officers in Los Angeles now hear this reminder often: "Remember, you're being videotaped!"

As Mary Melton wrote in *Los Angeles* magazine, "We look down on the car from a vantage that only God is customarily privy to. Whether we're in our living room or in front of a bank of flat screens at Circuit City, we, too, can stand in judgment of the accused. . . . Like God, we have a broad view of the picture and can see what lies ahead—Oh no, a cul-de-sac!—before the driver can."

Depending on the time of day, only minutes may pass before a flock of news-copters gathers, vulture-like, above any scene of trouble. Events too ephemeral and minor for a suburban newspaper's back pages can be more than enough to interrupt regular programming across the L.A. Basin, such as the day in 1997 when four helicopters

---

* Such copter-swarms do have rules. One rule is that the first news helicopter on the scene of a chase gets to take the ideal position, which is high and close. Later arrivals have to take up spots farther back or to the side. That means they must also fly lower to avoid having other helicopters in the way. It is analogous to reverse stadium seating.

followed a dolphin being trucked to a marine-mammal medical center. Striving for urgency, one pilot-reporter told viewers, "It's an ongoing rescue situation and we'll stay on the scene."

In the early years, news organizations were strongly inclined to join forces with law enforcement and firefighters, when duty called. During the Philadelphia riots in August 1963, the Go Patrol traffic helicopter sponsored by Atlantic Refining flew low over rooftops in riot zones and alerted the police about caches of rocks and bottles. John Carlton of the Go Patrol, upon seeing a fire at a Philadelphia hotel in 1968, landed on the roof, went down to the top floor, and began rousing occupants.

Bob Tur kept a tally of his rescues: sixty-seven people lifted out of danger and nine crashed airplanes found. Once he flew out to the Mojave Desert to locate a candidate for a replacement kidney; the recipient was camping at the time and out of touch—except by loudspeaker-equipped helicopter. His biggest single haul was at the Portofino Inn in Redondo Beach, California, during a 1988 storm. Hearing a report that fifty-four people were stranded on the roof, he flew his AStar helicopter to the scene and judged that twenty-two-foot waves were threatening the pier on which the restaurant perched. Tur dropped off a crew of firefighters and then shuttled the patrons to safety in twelve trips, buffeted by sixty-knot winds. "I always help people. I've done CPR on people on the street," says Tur. "I stop at accidents and wait for the paramedics."

"Bob's attitude is there is nothing we can't do with a helicopter," says Larry Welk, as in Lawrence Welk III, who flew with Tur as a cameraman before becoming a news pilot himself. "In a helicopter, he was a great stick." Welk has gotten into the action as well, once landing at the site of a fiery helicopter crash so that his cameraman could pull the occupants to safety, and another time while serving as cameraman on a helicopter that Tur owned.

Tur often went against the grain in his use of helicopters, including several imbroglios with the city of Los Angeles regarding where he could fly at emergency scenes. Tur is out of the news-pilot business now, but Welk is running his own operation, called Angel City Air. He invites me to pay a visit in February 2005. When I enter, a videotape from the previous night's broadcast is playing inside the Angel City office in a hangar on the north side of Whiteman Airport.

The tape shows a white ambulance, which has been stolen from Huntington Beach, streaking at night along a hilly, congested road, with police cars chasing it. The chase hits eighty miles per hour but ends when the ambulance loses control and crashes. Now, that was a good chase, the kibitzers of Angel City agree: it lasted long enough, it came during evening news, and, best of all, it was captured by Derek Bell, one of the news pilots employed with Angel City. Angel City is on contract with Channel 2 and Channel 9, CBS affiliates in Los Angeles.

Whiteman is a small airfield in the San Fernando Valley, on the north side of the Los Angeles Basin. The ambulance chase justifies some jubilation in the hangar because the "sweeps" period for February started yesterday. Still, television news directors have a short memory, particularly during sweeps: What will Angel City bring back today?

While there are no Emmys for covering car chases and most of the events are too minor to appear on any front pages, those in the business have nominees for great chases. There's the stolen 7-Up truck, whose street maneuvers sent hundreds of two-liter bottles across the streets; a stolen Greyhound bus dragging a chain-link fence; a stolen army tank; and the case of the Burning Lumber Truck of November 2001, in which the driver of a stolen eighteen-wheel truck drove around Dallas, Texas, for two hours, waving and honking his horn at pedestrians and news helicopters as the rear of the trailer belched smoke and sparks. In the early stages, the legal driver of the truck clung to the cab, which made for extra viewing thrills. It was good for forty minutes of nationally broadcast video.

Although some chases have ended in a shocking manner, the viewing public has made it clear through instant viewer ratings and TiVo reports that a good chase, captured live, is never wasted, regardless of whether it makes the newspaper the next day. Sometimes news-copters capture another helicopter making the news.[5] To be worthwhile, a chase needs to stay on the air at least twenty minutes so that viewers can tune in.

During sweeps, Angel City's two AStar helicopters are in the air eight hours per day, five days a week, reporting for two CBS affiliate stations. They are sure to obtain video of problems with L.A. traffic; they are sure to gather "beauty shots" for the opening and closing of each newscast. But mostly they fly the extra hours in the hope of catching some dramatic event that can be hurled into the ether before the competition arrives. The ideal crisis begins just before the local news programs, and is so riveting that viewers call their friends with those immortal words, "Turn-on-the-TV-quick!"

Bagging national-news-grade events takes persistence, much luck, and sometimes much nerve. Late on a hot afternoon in July 1986, Minnesota pilot Max Messmer saw an F2-strength tornado under way and pursued it for more than a half hour. Messmer edged as close as four hundred yards (sustaining minor damage to the aircraft) so the cameraman could broadcast a hummingbird's-eye view of the twister as it churned through five suburban cities and a nature refuge. The twister entirely preempted the station's 5:00 P.M. local news slot and was pipelined to network television. KARE-TV (previously ranked a limp third in Minneapolis ratings) received nationwide publicity and, thanks to its helicopter, became a force in the local TV market.

In 1997 one of Angel City's helicopters returned to base with a small badge of honor—a bullet hole in its tail boom—following coverage of the North Hollywood shootout for KCAL. In that event, two heavily armed and armored robbers tried (and failed) to break through a police cordon after robbing a bank. The broadcast even included a scene of gunman Larry Phillips turning to shoot at the camera.

The probability of sending home a comparable "money shot" during any given news broadcast is extremely small, but the reward to Welk's client station is enough to put him and a cameraman in the air, and at great expense. It's 5:00 P.M., one hour before the KCAL-TV newscast and time for Angel City to go to work. Welk gives the Sky 9 helicopter a quick once-over, climbs in, and starts flicking switches. This AStar model B2 costs well over a million dollars equipped, and burns thirty-eight gallons of jet fuel per hour. The manufacturer, Eurocopter, now sells more turbine helicopters in the United States than Sikorsky and Bell. Sky 9 has seats for three people plus a miniature control panel for Gil Leyvas, the camera and microwave-link operator. Larry is pilot and reporter. I will enjoy the view from the left seat, which is like perching on the edge of a magic carpet but without getting bugs in one's eyes. This AStar shows

7,000 hours on the hour meter, which for a car would be like having a quarter-million miles on the odometer.

Welk spools up the turbine and narrates the proceedings to me between radio calls to air-traffic control. Once the rotor has a firm grip on the air, we accelerate at a breathtaking pace, climbing out of Whiteman Airport as if on an escalator. This leaping effect is called translational lift. Translational lift provides a boost to any forward-flying helicopter and explains many seemingly outlandish events from the book of helicopter history.*

In 1975 Air America pilot Tony Coalson used all translational lift had to offer when faced with the need to evacuate nine fellow helicopter pilots from the roof of the USAID building in Saigon during the desperate hours before the city fell. Though his helicopter was so heavy it could barely budge from the helipad, no one offered to get out and make his job easier. Coalson lifted the collective lever briefly; the helicopter rose reluctantly into the air; then, before the Huey's rotor speed could slow unacceptably, he shoved the cyclic stick forward and dived off the roof, taking care to avoid striking the tail boom on the building's parapet. Profiting from translational lift gained during the fall, the Huey began flying before it hit the ground.

Welk is on patrol now, working his way toward South Central Los Angeles. His lifestyle could summon up envy in most working stiffs. The television-news contract is worth millions; and he wears what he wants to work, which today is khaki pants, a green short-sleeved shirt, and hiking shoes. He says the helicopter is just another article of clothing: "It's like a backpack to me. I put it on with each flight, like an army guy puts on his big sack." With this backpack he can vault over the worst traffic, at least while on the job. (He commutes to work in a car.)

Once upon a time, Welk was a profoundly unhappy theater major at San Diego International University. But flying made him happy. He began spending so much money on secret flying lessons that his father, Lawrence Welk, Jr., thought he was hooked on drugs.

* By bouncing along pavement on its skids, accelerating with each noisy bounce, a heavy-laden helicopter can sometimes gain enough translational lift to bootstrap itself into flight, even though a normal takeoff is impossible. This is called the "running takeoff" and was a common way for overloaded gunships to start the day in Vietnam.

Larry III dropped out of college and took a job fueling aircraft at the Santa Monica Airport, which is where he first encountered Bob Tur and wife, Marika. "When I was fueling their helicopter," Welk says, "you'd know if there was a story—their car would come screeching up and the doors would slam open and they would tear out to the helicopter." One day Tur said he needed a cameraman to fill in the next day and asked Welk if he knew of anyone. Welk said that, remarkably, he himself happened to be a cameraman.

Welk stretched the truth, and about as far as had Paul Cornu back in 1907. "So that night I went to my buddy who was a cameraman, and said, 'Look, I don't want to minimize what you do, but I need to know how to run a camera by tomorrow,'" Welk confesses. "Bob probably knew I wasn't a real cameraman but he also knew I was extremely motivated."

Welk went on to get his rotorcraft license and to set up his own helicopter charter company. He's making a living distinct from the rest of his family, without drawing from his grandfather's champagne-music empire. Each year Welk puts on goodwill parties that mix helicopter pilots, air-traffic controllers, police pilots, firefighter pilots, cameramen, fuelers, and news directors. A party at his house in December 2004 was easy for guests to find because a friend in the movie business mocked up the aftermath of a holiday helicopter crash in Welk's front yard, as if Santa Claus the chopper pilot had augered into the front yard and scattered a load of presents across the simulated snow. (Welk, never one to stint on a party, paid to have a totaled JetRanger hauled in for the display.) During the festivities, a police helicopter hovered overhead briefly to pay its respects.

We are over some kind of traffic-laden freeway when Leyvas advises Welk via the intercom that a bug has collided with the cover of the front-mounted camera. They need to set down so Leyvas can get out to wipe it down. This sounds easy—we're in a helicopter and helicopters can land almost anywhere—but that's theory. The political reality is different.

Even though Los Angeles sees dozens of helicopters active each weekday, antiheliport sentiment has been such that there are no helicopter gas stations in the area that sell services and parking, other than those at conventional airports. New York, of all places, is friendlier to helicopters, though Los Angeles was the original stronghold of helicopter mobility. Once Los Angeles had rooftop heliports for celebrities to use when flinging themselves over the heads of crowds, such as at the Millennium Biltmore Hotel (where the Beatles landed in 1964), or the Beverly Hilton (where John F. Kennedy landed to slip past civil-rights picketers in 1963). Other rooftops for public use were atop the Pacific Electric Building and the Ambassador Hotel. Dozens more were available at ground level.* A company called Helicabs began using the hotel heliports in 1961 for bringing passengers to the international airport. By 1962, Los Angeles was home port for one out of every five civilian helicopters in the United States.

The welcome mat has been rolled up, Welk says. "Los Angeles is so helicopter-unfriendly now. I can't believe there's no regular public heliport in downtown L.A.," referring to a facility that would offer parking, maintenance, and fuel. The city operates the Hooper Heliport, one of the world's largest rooftop heliports, featured in the movie *Blue Thunder,* but it is only for use by public-safety helicopters. Many of the smaller fixed-wing airports are gone also, having been sold to real-estate developers for much capital gain. But after a thoughtful moment Welk presses his intercom switch and adds: "There's so much wealth in a small area—if all these people were allowed to commute with helicopters, it'd be worse than *Apocalypse Now.*"

Welk radios for clearance into the Los Angeles International Airport and heads for a big H painted atop a parking garage. There are no other helicopters on the roof. On the way down, Welk explains why he lands at the far edge rather than the near edge: it extends his safety margin in case of engine failure on the way in.

"If you thought about all the ways this helicopter could kill you you'd never fly," Welk says as he settles the AStar on its skids, "but there's not a machine that gets more inspections, so that makes me feel good." Welk says that standing at the edge of a building roof makes him nervous, but looking down from his helicopter never does.

Welk takes us aloft again and gets the call to look in on a police

---

* A 1962 survey showed fifty-two heliports in the Los Angeles vicinity, plus sixteen conventional airports that received helicopters. Los Angeles Airways alone operated thirteen.

report about a stolen tractor: A suspect is driving a front-end loader down a busy street in El Monte near the I-10 Freeway and deputies are in hot pursuit, reaching speeds of twenty miles per hour. It's not something likely to make CNN, but Welk dutifully leans the ship forward with the cyclic. The turbine adds power and we reach a top speed of 113 knots. He slows over the tractor, orbits the machine and the police cars now around it, sends a quick televised report to the station, and resumes the sky patrol. Without seeming effort, via radio and intercom, over the next half hour he chats with a string of air-traffic controllers, Leyvas, me, other news pilots, his home base, and two or perhaps three people at KCAL.

"What the wife is saying is all white noise," Welk says, "but I can track six things in a helicopter."

He points out the scene of a train wreck that Angel Air shot two weeks earlier: "We're the eyes for the station. Maybe it cost nine hundred dollars to get that shot, but the station got two million viewers because of it." On the way back he swings through Topanga Canyon and shows me how to spot wire hazards by looking for poles and towers, which are more visible from the air than are wires.

We take to the sky again before 10:00 P.M. the same night, so Sky 9 can be in position to catch anything good that might happen concurrently with the nightly news. The air traffic hasn't let up; Welk estimates from the radio volume that at least fifteen helicopters are in the air. After transmitting a shot of heavy traffic on Interstate 405—it looks like white-hot lava flowing into a valley—Welk demonstrates the initial entry to an autorotation maneuver and spirals without power, expertly toward a tree-lined suburb.* "If this was real, we'd immediately establish a sixty-five-knot glide," he says as he returns power to the rotor.

Though the Los Angeles Basin by day is a choking brown haze of cars, parking lots, and building roofs, the scene at night is magical. The air is clear. The landscape gleams richly with colored lights and it wears strings of swimming pools that glow like aquamarines. One

---

* In an autorotation maneuver, the main rotor continues to turn in the same direction as it does when under power, driven by air flowing from underneath. This produces enough lift to brake the helicopter somewhat on the way down. A well-timed pull on the collective lever just prior to touchdown can add more braking power and can lead to a safe landing. Surprisingly, a freewheeling helicopter rotor is more effective in slowing a fall than a conventional parachute of the same diameter would be.

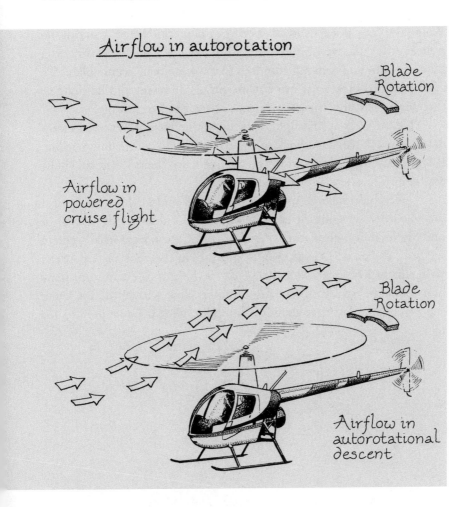

Airflow in autorotation

Blade Rotation

Airflow in powered cruise flight

Blade Rotation

Airflow in autorotational descent

Latin American pilot compared his nighttime work over the big cities of Brazil to flying inside a Christmas tree, and now I see what he means. We salute the Hooper Heliport, thoroughly framed with landing lights. We pass the Aon Center. Once it was called the First Interstate Bank Building and was the scene of a major fire in May 1988 that saw city helicopters used with great effectiveness for rooftop evacuations, for transporting firefighter crews, and for pinpointing the location of workers trapped in the building.[6]

On occasion, Welk says, he and Leyvas have seen suspects outrun or otherwise evade police officers and have radioed an alert. One man who slipped through the cordon stole a city bus. "We don't see ourselves as doing police work but just doing like anybody would,"

Welk says. "Say if a guy breaks into your neighbor's shed, you call the police. After all, we've got this really cool observer platform and a kick-ass camera. From five hundred feet we can read license plates." I ask Leyvas to pick a car out for me on a side street and he obliges, throwing the California tags on the monitor in front of me.

I leave Welk at midnight, locking up the hangar. It's been a long day, and Welk still has a stack of paperwork to do before going home. Regardless, his cheer about the airy life of a Southern California helicopter entrepreneur is unabated.

"We could have a surfer bit by a shark and in the same hour a snowboarder who hit a tree. Then a flash flood in the desert, all in one flight. If somebody keys Leonardo DiCaprio's car, that's where we go." Welk and Leyvas keep a "go bag" in the back of the AStar, with survival rations and a change of clothes, in case a mighty temblor rends the San Andreas Fault asunder. That will be the Big One, the legendary day that draws every pilot with a helicopter to his name.

Chapter Fifteen
The Great Stick

When helicopter pilots get together to tell stories, a term they often use for a worthy pilot is a good stick. That's someone who has gone beyond competency with the controls and systems, and who has good judgment and a good hand with cyclic and collective. And that elite corps itself has a top echelon—call them the Great Sticks—who have shown the community of pilots and owners how to use helicopters in new ways. These are the people who do the previously impossible. What's more, they come back alive. The qualities that a Great Stick possesses are probably ineffable, but I am told they include finesse on the controls, complete knowledge of engine and systems without gap or pause, and a clear mind to read the situation and weigh risks and benefits.

Piloting finesse is a quality more obvious by its absence than by its presence. Each October the American Helicopter Museum in West Chester, Pennsylvania, puts on a weekend air show. The Rotorfest offers visitors the chance to ride in a JetRanger, to climb inside military transports, to look at war memorabilia, and to walk through the permanent historical exhibits in the museum building. One hands-on exhibit at the time I visited was a bobtailed version of a surplus Schweizer two-seater helicopter, with its main rotor blades and tail boom shortened so that it fit on a small trailer. An electric motor turned the rotors at a carousel pace. The collective lever, the cyclic stick, and the pedals were all connected to the blades, so a patron could see how stepping on the left pedal caused the rotor blades on the tail to steepen their pitch and bite the air more deeply. But the kids on the display were having none of such detail: they were yanking on the levers and stomping on the pedals with vigor.

By contrast, working pilots handle the controls with subtle regard, moving instantly but barely, so that backseaters may wonder if the pilot is doing anything up there. Five years earlier I rode in a small helicopter while an electrician on board worked on a live transmission line in Pennsylvania. We were eighty feet off the ground, and pilot Mark Campolong had to hold his machine next to a thumb-thick aluminum-steel cable carrying 230,000 volts of electricity. His job, simply stated, was to keep electrician Jeff Pigott close enough to the cable to do his work, but not so close as to tangle his ship with the line. New pilots need a football field or larger when learning to maneuver; Campolong had no more than sixteen inches of tolerable error.* I watched his gloved hands cope with the light and variable winds: his hand on the cyclic was as economical of motion as a bicyclist who is cruising down the street.** After we banked and flew back to the fueling truck, he said that his mother-in-law asked why his work was so tiring. In her opinion, all he did was sit around all day.

Pilots at work *do* sit around all day, but some take those chairs to entirely new places, such as USAF Major Harry Dunn, who in 1965 first demonstrated the practicability of midair refueling by bringing a mock refueling probe on his H-3 helicopter to mate with a drogue trailing from a C-130 aircraft off Cherry Point, North Carolina. Another benchmark in piloting can be seen in the motion picture

* In such a situation, the pilot is conscious of two risks in particular: an engine failure or accidentally bringing the tail rotor against the cable. Either would lead to a crash.

** Experienced pilots tell me that once the abilities and instincts are fully developed, the mechanical skills of flying a specific helicopter are as permanent as the knack of riding a bicycle.

Hiller 12E

Terminator 2. During a nighttime chase scene, a Bell JetRanger dips down—bringing its skids nearly to the pavement—to fly under an overpass. It was director James Cameron's idea. Veteran pilot Chuck Tamburro put his helicopter on wheels and rolled it under the bridge to measure the clearance (five feet above and four feet on each side). He flew the stunt twice at a speed of sixty knots. No special effects were used.*

Another daunting task, appropriate only for Great Sticks, is landing a helicopter on slopes and pinnacles at altitudes that are at the very fringe of its capabilities. In May 1960 this was a job that Link Luckett of Anchorage, Alaska, took on successfully. For a few weeks it made his little red Hiller 12E Raven the most famous helicopter in the world. All fame is fleeting, but even so, decades later, the incident in Alaska at 17,200 feet says much about piloting, and it raised expectations among mountaineers worldwide about their prospects of coming back from death's door.

The story began on Mount McKinley in Alaska on May 17, 1960. Four climbers were involved. One was a relative newcomer to mountaineering: Oregon rancher John Day, who in 1957 decided he would set records by ascending mountains faster than anyone else.

---

* Tamburro has a cameo appearance as a police pilot, during a prevous scene in which the villain tells him to jump out of a hovering helicopter.

With Day on McKinley were three highly experienced climbers: Pete Schoening and the brothers Lou and Jim Whittaker. Getting a head start by arriving via airplane at 10,000 feet, the group had virtually speed-walked to the summit at 20,320 feet in three days.

They ran into trouble on the western side on the way down, two thousand feet below the top, when one climber slipped on the ice and the others didn't react quickly enough. Many climbing accidents happen in this manner during the descent, when climbers' reactions are slowed by oxygen deprivation, fatigue, and cold, and when downhill momentum is working against good traction. The four were linked by rope and the first man to go dragged the rest down a steep slope in a tangle of ropes, packs, and dagger-sharp ice axes. They stopped 400 feet below, where the slope flattened. That saved them from scooting off a lip and down a much steeper drop of at least two thousand feet. John Day had a broken leg and torn ligaments. Schoening had a concussion and in the coming days would slip in and out of consciousness and suffer increasingly from frostbite. The Whittakers were injured but mobile. Another climbing party from Anchorage saw the aftermath and lent them assistance but could not bring them down. Day could not move, so they slipped him into a sleeping bag and tent where he fell. And the Anchorage climbers who came to help were having trouble with another climber in their party: Helga Bading was critically ill with altitude sickness at 16,400 feet and couldn't move.

Day's biding place on the slope at 17,200 feet was too steep for an airplane to land. The army's H-21 Flying Banana helicopter tried to reach that altitude but could not manage it even at cruise speed. No other helicopters available in Alaska had the performance to reach the spot either, at least according to the published specifications. The French Alouette II was turbine-powered and had made mountain rescues at 13,000 feet in 1957; it would have been a good ship in which to try a rescue, but no Alouette was available on such short notice. Seemingly, the best any aircraft could do was to drop supplies. Even to approach the area was a real hazard. On May 20 a light plane flown by an Anchorage contractor came to Day's vicinity with plans to drop a set of spare radio batteries, but stalled while turning. That fiery crash killed the pilot and passenger instantly.

In Anchorage, a helicopter pilot named Link Luckett, age

thirty-two and glaringly bald, was following the crisis by radio. Luckett had begun flying a light plane in 1946 while working for his family's earthmoving business in Springdale, Arkansas. He went into the army, trained as a pilot, and later taught helicopter pilots at Fort Rucker in the earliest years of that program. After leaving the army he worked as a commercial pilot in Guatemala and on the Gulf Coast. He moved to Alaska in 1958.

"Link was charismatic," pilot Ken Moon says. Moon worked with Luckett beginning in 1960. "To be a bald guy was very unusual. At that time, the only other bald guy in the universe was Yul Brynner. He was a handsome dude, popular with the ladies. Most of all, Link was a man who lived for adventure."

On typical jobs Luckett used his Raven helicopter to haul people and supplies to remote areas, often as a flying crane to lift gear into place at remote construction projects. Once Hill-I-Copter flew a trapeze and acrobats for an outdoor circus performance.

Luckett's Raven was grounded when the first news came, but the replacement part that Luckett needed arrived at the Anchorage airport on the morning of May 20. That morning Luckett received two phone calls, one from Bading's husband and one from an army helicopter pilot, asking if Luckett could retrieve Helga Bading with his Raven, a single-main-rotor helicopter. Luckett responded that he would give it a try, but first he wanted the Anchorage party to carry Helga down to a decent landing spot at 14,500 feet.

Earlier models of the piston-powered Raven were commonly understood as confined to altitudes of 5,000 feet or less. Planning on reaching an altitude three times higher might seem the act of a bold pilot rather than one who was intent on growing old, but Luckett specialized in mountain work and felt that aerodynamics and the chilly weather were in his favor. He had already taken this higher-performance E model Raven in a climbing hover above 15,000 feet as part of a manufacturer's test program, and he expected that the cold air and a strong wind would boost its performance sufficiently for a landing at 14,500.*

When the repair was finished Luckett flew north to Talkeetna, donned a borrowed set of cold-weather boots and clothes, and

---

* Cold air is denser than warm air, so rotor blades get a better grip on the atmosphere on a cold day.

proceeded to the mountain. He saw from the air that airplanes had already landed at 14,500 feet and were prepared to bring Bading out. Dozens of climbers were gathering there to ready a climb that would assist the men still trapped at higher elevations. Luckett landed at the spot to consult. Glacier pilot Don Sheldon walked over and pointed to an overhanging curl of snow atop a vertical wall three thousand feet higher: "They're up there," Sheldon told Luckett, meaning the injured men from the Day climbing party. "Go get 'em and bring 'em down here, and I'll take 'em to Anchorage." Luckett agreed to scout out the situation and, breathing oxygen through a tube from a tank that the air force had loaned him, took off in a climbing circle. Luckett was pleased by the Raven's performance: it passed between the twin peaks of Mount McKinley, which meant it had reached 19,500 feet.

The injured John Day heard it first: a faint whopping noise that echoed off the rock faces. A small helicopter painted in fluorescent red was coming in. It swung past the wrecked plane and aimed downhill to make a running landing. With its nose high and the tail-boom stinger plowing a furrow in the snow, it turned sideways and skidded to a stop short of an abrupt dropoff. Leaving the Raven running, a man in an air force bomber suit and aviators' boots climbed out and slogged toward the still-smoking airplane wreck. Then the pilot fell face-first in the snow.

Luckett, who lived in Anchorage near sea level, was not ready to handle three and a half miles of altitude with no acclimation. But the snow revived him and he lumbered back to the helicopter for another whiff of oxygen. As Luckett came back to full consciousness, the Whittaker brothers decided that Luckett had come for Day, so they helpfully lugged their friend toward the helicopter, walking close to the main rotor blades on the uphill side. Those blades were spinning at lethal speed and were not easy to see.

Luckett saw the Whittakers' close shave and jumped out. By yelling and gesturing over the noise, Luckett explained that this was just a test to see whether a rescue would be possible later. The helicopter might not produce enough lift to free itself from the snow and get moving forward; or once sliding it might simply skid out of control before it could get safely aloft. He warned them about the tail rotor, which rotates like a vertical set of unshielded lawn-mower blades. With the Whittakers steadying the Raven from slipping sideways as he turned it to face downhill, Luckett's helicopter started sliding down the slope, where it could take advantage of translational lift.

Luckett took off safely and parked at the rescue camp at 10,500 feet. His radio message went uphill: The Whittakers should call back when winds were at least fifteen knots, because Luckett would need all that wind to get off the snow with the extra weight of a passenger. But the decision to wait also posed a risk. If winds did not rise until evening, he would lose the sun-warmed air that flowed up the west side each sunny afternoon, and also would give up the lighting necessary for good depth perception.

Winds suited five hours later. With the help of air force mechanics and the Hiller factory representative, Luckett lightened the Raven with a vengeance. They drained the fuel tank and put back only enough fuel for a half hour of operation. Luckett started the gasoline engine and they removed the thirty-five-pound battery, using two lantern batteries to run the instruments. Without a battery there was no need for the starter motor, so they took that out. And without a large battery the high-frequency radio wouldn't work, so that came out also. Though the temperature was twenty below zero Fahrenheit, they took off the doors.

After one aborted landing in deep snow and an emergency takeoff to avoid getting mired, Luckett flew around again and headed toward a rock face called the West Wall, where he used the steeply sloping ground like a brake. He spun the machine around and headed toward the Whittakers. Okay, he signaled: now it was time to load Day aboard.

Luckett crossed his gloved fingers at his passenger as if welcoming any good-luck wishes, then raised the collective slowly and began scooting down the slope. Even with the aggressive removal of all optional weight, at high altitude Luckett's Raven entirely lacked the distinctive feature that helicopters usually have over airplanes, which is their ability to hover. The air was so thin that the helicopter could not hang in the air unless the airspeed needle showed at least fifteen knots.[1] According to the instruments, his Lycoming gasoline-fueled engine at an altitude at 17,200 feet was producing only slightly more power than one idling at sea level, so a very fine touch on the controls was called for.* Luckett lifted off the slope and brought Day to a landing strip for transfer to an airplane that flew him to Anchorage. He came back to pick up Schoening the next morning. The Whittakers were able to walk down to the landing zone at 14,500 feet; hours later a storm closed down more flying. When Luckett contemplated one more trip later to recover the body of the pilot killed in the airplane crash on May 20, his insurance agent informed him that a special premium would be charged for the attempt. Climbers retrieved the body instead. Luckett insisted that there was nothing death-defying about the rescue mission since he knew the Raven was a good high-altitude performer, and the extremely cold air added just enough to its performance to make it all work. But others thought it was edgy indeed. Luckett appeared on national television quiz shows and received a Carnegie Medal for heroism.

"Link had the world by the butt for a time," says Ken Moon, who was on his way to Alaska at the time to serve as a second pilot for the company. "Everybody who needed a pilot came to Hill-I-Copter," Moon says.

What mix of experience goes into making a Great Stick? During Luckett's generation, an expert pilot earned that reputation by excelling across a wide variety of lessons and jobs (some of which are considered too dangerous for pilots today). Such a Greatest Generation résumé might list the following: Training under world-class instructors at Fort Wolters and Fort Rucker.** Mastering emergency autorotation landings over many thousands of practice sessions.[2] Lifting heavy loads at high altitude. Doing "toe-in" landings on mountainsides. Flying through smoke and dropping hundreds of gallons of water with precision onto wildfires. Meeting the

---

* This is called power management. In such a situation excessive use of the collective lever (which by steepening blade pitch raises aerodynamic drag) will slow the rotor to dangerously low speeds.

** From 1956 through 1973, Fort Wolters in Texas trained 40,000 pilots, ranking from warrant officer through general officer. During the peak of the Vietnam War, instructors at Wolters typically fell into two groups: warrant officers with combat flying experience in Southeast Asia, or civilians trained and employed by Southern Airways, a contractor. Hugh Mills, later an Aero Scout pilot in Vietnam, was trained as a lieutenant by Southern Airways. The training was rigorous and intense and required many practice autorotations in case of engine failure. Even so, some pilots found the opportunity during solo flights to engage in extracurricular activities that included playing helicopter tag near Possum Kingdom Lake and picking up girlfriends at remote training sites.

special tests that must be passed to haul humans at the end of a cable. Flying to medical emergencies while having the guts to turn back if conditions are unsafe. And, of course, hours and hours of combat flying in Vietnam. Fairly or unfairly, for nearly four decades this has been the gold standard of expertise for any pilot born around 1950 or before, like Luckett or Chuck Tamburro.

I located another Great Stick by asking an old hand with the 160th Special Operations Aviation Regiment, Clifton O'Brien, which pilots from Vietnam still stand as an inspiration to the unit. One he named without hesitation was Justin G. "Guy" Ballou, Aero Scout pilot.

First, the historical setting: In Vietnam, the infantry was supposed to make contact with enemy soldiers and fix them in position so that the full firepower of American aircraft and artillery could be brought to bear. As the commander of the Eleventh Armored Cavalry Regiment, General George S. Patton III, put it, his men were to "find the bastards, then pile on." One way to find the enemy was to send out light and nimble armed helicopters called Aero Scouts.

According to Patton, one of the most effective Aero Scout pilots in the regiment was Guy Ballou of the Eleventh Air Cavalry Troop. Ballou did his scouting chiefly across the III Corps zone of central Vietnam. The principal task of Ballou and his gunner-observer was to sniff out traces of enemy habitation and—this is the tricky part—determine whether enough enemies were on hand to justify a full-scale attack. At this time, 1968 and 1969, "sufficient" meant at least thirty soldiers. Aero Scouting had to be done below treetop level and quite close to hidden enemy machine-gun bunkers, and therefore was awesomely dangerous.

"The key to finding the actual base camps was looking for a straight line," Ballou says now, "as in nature you don't find them." But finding a camp was not enough either. Only a few caretakers might be present, and this was not worth calling in the infantry and gunships. Or the camp might have been abandoned. If the site looked promising but signs suggested no one was home, Ballou and an observer made a note of the location and would drop by occasionally for signs that troops had taken up residence. He was careful to vary the direction and timing of his passing so as not to tip off the enemy that the campsite was under surveillance. The daily challenge was to fly lowly and slowly enough that he and his observer, Robert Hepler, could see what they needed to see without getting shot down.

Ballou flew an OH-6 Light Observation Helicopter, nicknamed the Loach for its initials. The crew's greatest protection lay in the machine's speed and agility rather than armor plate. The machine was only twenty-seven feet long and extremely maneuverable, and therefore was harder for gunners to hit than a Huey.[3]

The Loach "fit like a glove and flew like a hummingbird," according to Ballou's observer and gunner, Robert Hepler. "It could almost take whatever Charlie could throw at her, with the exception of that one day where we went through three ships." Those expended helicopters, numbered 331 through 333, had been new and in his platoon only a few days before Vietcong fire totaled them. "I can think of few days when we were not in a firefight," Ballou says.

Though standard Aero Scout helicopters carried a Gatling-style minigun, Ballou regarded it as heavy and troublesome and had it pulled out in favor of his personal favorite, M-60 light machine guns: one for Hepler in the left front seat, one for an observer in the backseat, and sometimes a third one for Ballou in the right seat. At such times Ballou gripped the cyclic with his left hand, nudged the collective lever with his knee, and guided the gun with his right hand.

Hughes OH-6 Loach

Hepler also favored a forty-millimeter grenade launcher and brought handmade bombs along to drop on bunkers. "We would load the Loach till it could hardly get off the ground," Ballou recalls. There were times when, out of ammunition but hot in pursuit, Ballou pursued enemy soldiers across clearings and ended the chase by landing on them.

Enemy main-force soldiers were masters of concealment, moving by night and staying under cover all day, so a helicopter might seem an unlikely instrument to find any of them. But Ballou had a secret weapon. He went up each morning at daybreak and headed for clearings that he knew were not occupied or patrolled by American or ARVN troops. Then he used the glancing light of morning like a sort of backscatter radar. He was looking for signs of what he calls "shininess" in the grass. Shiny grass indicated that people had walked across it during the night. Smells of cooking and human waste also pointed to current occupation. So did stretches of muddy stream. General Patton once credited Ballou for pinpointing the location of the Dong Ngai Regiment in 1968. In passing through the area, Ballou was at his usual low altitude. How low? So low that Ballou noticed fish were swimming around in flooded bomb craters. This was far from a river, and Ballou surmised that someone must have put the fish there as a food supply. This was Aero Scouting at its best.*

And it is certainly the kind of thing that Great Sticks do. But it is also ancient history, given that today's pilots aren't learning under Vietnam War conditions. Are Great Sticks still being born and made somewhere out there? Or will retirements among the Vietnam-trained generation—once a trickle but now an exodus—exhaust the population forever?

Two hours spent with the imperturbable Bob Steinbrunn in his rooftop aerie offers me a chance to check in on the state of helicopter piloting, then and now. Steinbrunn has the credentials to make the comparison. When I visit him at his workplace in Minneapolis, Minnesota, he has thirty-nine years of experience and his logbooks show 14,000 hours, most of which is helicopter time, including years

---

* In May 1970, another Aero Scout helicopter in Cambodia found subtle signs that led to a giant weapons cache, dubbed Rock Island East, bulging with 7 million rounds of ammunition and thousands of rockets.

flying loads for construction projects.⁴ He wears a uniform similar to
that of an airline pilot: white shirt with epaulets, and black slacks.
His hair is short and blond; he speaks with calm authority. The air
ambulance operation for which he flies started moving patients in
1985 and has operated with zero crashes or incidents. Steinbrunn was
here from the first day and still loves his work. There are four pilots.

"We have the biggest picture window in the world," Steinbrunn
says, showing me to a chair. He's referring to the extra-wide door of
this rooftop hangar at North Memorial Hospital. We can see downtown Minneapolis from this elevation. As with all heliports the view
is unobstructed by a fence or railing, since these pose a hazard to helicopters. Thirty feet from his desk is his other chair, the pilot's seat of
an Agusta 109 medical helicopter.

Steinbrunn entered the army in 1966 and trained as a helicopter
pilot, then flew a pink-team gunship to provide cover for Loach helicopters with the Seventeenth Cavalry in Vietnam. After his year in
country he went back to teach at the army's helicopter school. Out
of the army, he flew a variety of aircraft before becoming a medical
helicopter pilot in 1985. On such flights the crew numbers three:
Steinbrunn, a flight nurse, and a flight paramedic.

"I need to know this ship so well because I have to find all knobs
by touch," Steinbrunn says. "Say I'm at a car on fire and I've got wires
and trees. I don't have time to do more than glance at the panels.
Then it's landing gear down and locked, parking brake on, torque
matched on both, and crew belts fastened. I have a minute or two for
all that." He flies without a copilot, and so handles all his own computerized navigation. Even with the occasional difficulties, he says,
the toughest day he spends as a medical pilot has never approached
the stress and hazards of flying in Vietnam.

"This kind of work is childishly simple by comparison,"
Steinbrunn says. "In Vietnam if there was an American unit on the
ground, if you didn't get there in time, they were toast, so you did
everything conceivable to get them out, including killing yourself."

Later I ask Colonel Greg Lengyel, a new-generation helicopter
pilot for the air force, about the direction of piloting in the post-
post-Vietnam era. He begins with an observation of pilots old and
new, from his training days: "If there was a style difference, I'd say
that the Vietnam guys were so experienced that it was virtually im-

possible to rattle them in the cockpit," Lengyel wrote back in an e-mail. "They were more 'stick and rudder' guys than they were 'technical knowledge' guys."

Today's helicopters need different skills, according to former Black Hawk pilot Craig Dyer: "The 'around-the-clock' high-tech capabilities are much more comprehensive and demanding."

"Perhaps we are creating another generation of pilots that will be so combat seasoned that they too will be cool under almost all circumstances," Lengyel believes. "I know that many of the young MH-53M pilots right now have more combat time than time flying in the CONUS."* He flies the MH-53J/M Pave Low, which is customized for delivering special-operations forces without detection.

For seasoned military pilots wanting to move into special-forces work, one of the very few graduate-level courses in the world can be found at Fort Campbell, Kentucky. Its humble origin can be seen in the southwest corner of the post. It is a tan, concrete-walled building. It is off the main roads and surrounded by tree-covered hills. A sign indicates that the building with the bomb casings out front is the headquarters for an explosives-disposal unit, the 717th Ordnance Company. In 1981 the building had a different identity: It housed the super-secret Task Force 160, the helicopter unit preparing for a second try at spiriting fifty-three American hostages from hostile Iran.

The reason for Task Force 160 and its fleet of OH-6, Vietnam-era helicopters was the humiliating failure of an earlier helicopter-dependent rescue attempt in 1980. It was code-named Operation Eagle Claw and was popularly known as the Desert One mission. The origin of Eagle Claw was a time of high tension between the United States and the new Islamic regime over the fate of Shah Reza Pahlavi, the deposed ruler. Less than two weeks after the United States allowed the shah to enter the country for surgery, armed revolutionaries broke into the U.S. embassy in Tehran and took dozens of prisoners.

A few Americans were released, but that still left fifty-three hostages. President Jimmy Carter authorized an elaborate plan in April 1980 in which forces from all branches of the armed services would go deep into Iran to retrieve them. As part of this, eight navy

---

* Continental United States.

RH-53 Sea Stallion helicopters, normally used for minesweeping, would depart from an aircraft carrier and fly 600 miles to a remote landing strip called Desert One. Forty-four aircraft would participate in six elaborate phases.

Things started going wrong early, when three of eight helicopters dropped out with real or suspected mechanical problems. That violated minimums and the rescue was scrubbed. Still, the story was not over. While one helicopter was hovering in a dust cloud, its pilot struggling to maintain his visual fix on the ground, its rotor blades cut into a C-130 transport. The shrapnel ignited fuel and ammunition on both aircraft. At that point it was a scramble for survival. In the departure, the task force left eight bodies and much wreckage behind.

Eagle Claw generated many official recriminations and reports about how to avoid another such debacle. One was a statement by special forces that they needed a helicopter-like vehicle but with greater speed and range, which in time led to the V-22 Osprey. The lessons ranged from the grand scale (the United States had let its conventional military fall apart after Vietnam) to the minute (the apparent warning on one helicopter that a blade was about to fail actually didn't indicate any threat to safety during the mission). The least publicized response was a plan, code-named Honey Badger, to go in one more time with more helicopters and commandos, but different tactics.

For Honey Badger the army gathered new H-60 Black Hawks, old Loach helicopters, and troops from the 101st Airborne. The task would be enormously difficult. Forewarned now, the Iranians had dispersed the hostages across multiple locations and also kept them moving.

The MH-6s trained by night, and evaded notice during daylight hours by sliding into garage stalls in the back of the building each dawn. They were leftover Loach helicopters from the Vietnam War, considered by many to be obsolete after the introduction of the Black Hawk. The army decided differently and rounded them up from National Guard units. "The Huey and OH-58 didn't suit," recalls Clif O'Brien, a retired command sergeant-major and a participant in the preparations. "The MH-6 is easier to work on, rapidly deployable, and crash-worthy. The deployability is excellent. You can put six MH-6s off a C-141 [transport], and in six or seven minutes you can have them running." The troops of SOAR call the MH-6 and its armed cousin, the AH-6, the Little Bird.

Preparations for Honey Badger stood down after the release of all hostages in 1981, but the army saw more asymmetric warfare in its future and decided to nurture a permanent, commando-style helicopter force.* The existence of Task Force 160 (now the 160th Special Operations Aviation Regiment) remained secret until its cover was blown during Persian Gulf operations in 1987. By then army helicopter pilots had come to regard it as the prime billet among combat units.

Nighttime training accidents were the first obstacle. After four helicopters crashed during training in 1983, a regimen was set up to train pilots about navigating and approaching a target with the early model of night-vision goggles then available; these were originally intended for use by truck drivers and had a narrow field of view. Task Force 160 did battle for the first time in the island nation of Grenada, supporting an American action to oppose a Marxist movement that was receiving Cuban support and to evacuate American citizens. Instead of its preferred night approach, however, the helicopters had to come in during the day. They took heavy antiaircraft fire on approaching one target, a prison. One helicopter crashed, for the unit's first combat fatality. Later the unit shot up Iranian gunboats and a minelayer in the Persian Gulf, used two of its MH-47 Chinooks to haul back a Soviet Mi-24 helicopter gunship abandoned in Chad, and fought troops loyal to General Manuel Noriega in Panama.

The latter action led to SOAR's second and third combat fatalities, when an armed Little Bird was shot down with a rocket-propelled grenade at the Colon harbor. The helicopter had been covering the exit of SEALs following a commando raid on "high value targets" at a beach house.

The reclusive regiment became headline news on October 3, 1993, because of a battle in the narrow streets of Mogadishu, Somalia. That afternoon a U.S. special forces raid arrived via SOAR helicopters at a building across from the Olympic Hotel. The action

---

* Many other nations have reached the same conclusion. As one example, in December 1969 Israel flew commandos in two CH-53 transports to an Egyptian outpost at Ras Ghareb, where they loaded up seven tons of components from a Soviet-supplied radar station and hauled them back for study.

Mil Mi-24 Hind

initially captured two dozen of Mohammed Farah Aidid's assistants but went bad after a rocket-propelled grenade hit the tail of one orbiting Black Hawk, which spun out of control and crashed a few hundred yards away. A second Black Hawk caught an RPG round in the cockpit and crashed a mile farther off. The battle to recover bodies and wounded men lasted well into dark, then resumed before dawn. In one of the most dramatic moments, an MH-6 Little Bird helicopter with the code name of Star 41 made a high-risk landing in a narrow street in the midst of the gun battle, gathered up Sergeants Daniel Busch and Jim Smith from the crashed Super 61 Black Hawk, and launched safely. The toll for SOAR aviators was five killed and one captured.

The copilot on Star 41, the Little Bird that touched down in the Mogadishu alleyway, was Chief Warrant Officer 3 Karl Maier. In warrior style, Maier makes no claim to heroism that day: "We were unarmed and afraid," he says, noting that the combined action of all the gunfire from armed Little Birds overhead was so fearsome that the Somali fighters stayed back and made the rescue possible.

Maier's current job at Fort Campbell is operations officer for the training arm of the SOAR unit, known to pilots and crew members as the Green Platoon. Though pilots enter the Green Platoon already qualified to fly army helicopters, the process to ripen them for SOAR work takes three months, followed by two years of additional work and instruction for those who want the authority to plan and lead a mission. Trainees in Green Platoon stay very busy. After two weeks of individual combat training, pilots spend three weeks planning and flying low-level night missions to unmarked landing zones at least sixty miles from the base. All navigation must be done with map, clock, and compass; no other gadgets are permitted.

"The first night, they get lost like last year's Easter egg hunt," says Maier. "They get real quiet suddenly, and begin looking behind 'em. As an instructor I'd say, 'Look at the big picture. What town could that possibly be?' But in three weeks, I'm not talking and they're hitting the target." Six to nine months of training prepares a newcomer to fly missions as copilot. The Green Platoon handles at least 120 pilots in a typical year.

Maier says he has no opinion as to which helicopter is better than another. He explains that because of high-altitude missions in Afghanistan, stalwarts like the CH-47 are getting more use than ever. Still, the Little Bird seems to come up spontaneously as we talk, probably because of Maier's time in Mogadishu and maybe because of how small wars have been playing out in urban settings, the most dangerous and frustrating of all battlefields for American troops. The MH-6 and AH-6 helicopters now used in battle by the SOAR pilots are modified MD530F models, rather than refurbished aircraft from Vietnam. The MH-6 helicopter weighs 2,100 pounds empty; fully loaded and fueled it weighs more than twice as much. Four fully armed soldiers can ride on the outside of a Little Bird, seated on fold-down planks attached to the landing-skid struts.

"The guys in the MH-6, they work close in, up to the front doors and to the top of the building," says Maier. "It's very good at urban warfare. Compare that to regular army aviation—their urban guys get you to the outside of town. They're not dropping you off downtown." The Little Bird is preferred for dropping off troops be-cause it is small and nimble, which makes it hard to hit from the ground. "They [the enemy fighters] don't know where we're going to land, and at night we're all blacked out," Maier says. "They're shooting at the noise and that's behind us."

After graduating from the Green Platoon, pilots selected for the AH-6 gunship learn to use the trusty, 2.75-inch folding-fin rocket.

"This is direct fire on a target, not standoff like the Apache [helicopter]," Maier says. "The good guys identify themselves and you shoot around 'em." At a distance of 200 yards Little Bird pilots can put a full load of rockets into a standard garage door, he says.

By the time an AH-6 Little Bird pilot is ready to graduate from Fort Campbell, he has had the benefit of learning from the world's best helicopter instructors. He has fired a small mountain of rockets, and a truckload of machine-gun rounds. From Fort Rucker onward he has run a tab well north of a million dollars for helicopter time.

Granting that such a curriculum and many missions will in time lead to a new generation of Great Sticks with a new set of skills, what opportunity remains for pilots not pursuing the military route? Winning the basic stack of commercial helicopter licenses costs as much as going to a good state college. The path of Dan Rudert into pilot-hood suggests that opportunity remains for private pilots, if they are sufficiently motivated. Rudert found his way into the upper ranks of helicopter pilots without the benefit of a wealthy family or of military helicopter time.

In 1978, at age nineteen, Rudert was a surgical orderly for a hospital in Salt Lake City. His life changed when one of the medical helicopter pilots let him ride along on a refueling flight and gave him a demonstration of combat flight techniques on the way to the international airport. The pilot aimed for the Utah State Capitol Dome, Rudert recalls, "put the collective to the floor, then pulled it up at the last minute. Stuff was floating in the air.* He was doing S turns—he just dove it out of the sky. We landed, and I just said, 'Whoa, what is this!'" Rudert hocked his Jeep CJ-5 to get enough money to start training. He keeps a picture of himself sitting on the couch, with his wife and five thousand dollars in cash. It would take much more than that amount of money to get his commercial rating, and much time after that driving a fuel truck in exchange for blade time, but now Rudert, 47, is a top-flight pilot who spends half of his work year flying in the elite world of moviemaking, and the other half flying among the Rocky Mountains to fight fires and make rescues.

Whatever became of the first generation of Great Sticks? John

* Without endorsing such stunt flying over downtown Salt Lake, Rudert explains that this zero-G maneuver was technically a side-slip, often called a hammerhead stall.

McElhinney, veteran pilot for L.A. radio news station KMPC and the Los Angeles Police Department,[5] always subscribed to the familiar saw: There Are Old Pilots and There Are Bold Pilots, but There Are No Old, Bold Pilots. Luckett was bold. Surely this truism removed Luckett from the gene pool, in some fiery crash on a disputed mountainside.

But no: Luckett survived his McKinley ascent, all his other work in Alaska,* his years in the South Pacific flying heavy loads of oil-drilling gear across the jungles of New Guinea, and ten years as a pilot for Air America in Laos and Vietnam. The latter duty included rescues and resupply missions into many disputed zones, acting as someone else's clay pigeon. Luckett cheated death through it all, and retired to a verdant hillside in Indonesia, where he lives today.

---

* Such as delivering supplies to mountaintop microwave relay stations while they were shrouded in fog. Luckett coped by hovering uphill at low speeds within a few dozen feet of the slope, so as to keep his bearings.

Chapter Sixteen
Conclusion

In 1953 a Rhode Island lawyer by the name of Leo Connors came upon a mysterious back stairway in his office. He had been told years before that his floor was at top of the Fleet National Bank Building in Providence, but now he could see that the building had some kind of bonus floor. Connors climbed the stairs and found a dirigible-passenger waiting room, built in 1928 and forgotten. It had been built before the explosion of the dirigible *Hindenburg* and the permanent end that the disaster brought to such travel. The abandoned furnishings, gray with dust, had the appearance of set dressings from the motion picture *20,000 Leagues Under the Sea*. There were red leather chairs and sofas, Art Deco light fixtures, fine brass fittings, oak paneling, a liquor cabinet, and a galley. The Providence waiting room must have looked to Connors like an artifact from some discarded timeline, an alternate past in which birds learned to share the rooftops with blimps and zeppelins.

Early helicoptrians expected that dirigibles' claim on the roof environment would be short, and they were right about that. But they expected that helicopters would take over the roost, and they were wrong about that. Even in Southern California, which has long been a hotspot for helicopters, and which has plenty of millionaires and billionaires able to afford them, there are no residential rooftop heliports.

As a seven-year-old I saw our destiny vividly. It was on display at the World of Tomorrow at the Seattle World's Fair. In that 1962 exhibition, my family went up a Plexiglas-walled elevator through a cloud-like mass of aluminum cubes meant to symbolize the future. We viewed the future's spinning, solar-powered home and its kitchen. Following that, a tiny helicopter lifted off from a home helipad and pointed the way to another diorama showing how people would get around the world of tomorrow. Workers in 2000 would commute by monorail, scoot along high-speed automated highways in turbine-powered autos, or fly to work via their personal "gyrocopter."

Instead we have today's helicopters that are powerful and capable, routinely accomplishing feats that even the most optimistic visionaries didn't forecast, but that cost a great deal. And despite all the publicists who compared them in ease of operation to cars, the real ones all demand skill and cool judgment from their pilots.

This book started out with the idea of the helicopter as a god machine, comparable to the Greek mechane. Opinions on helicopters' virtues and vices vary. Depending on the news of the day, a typical person might view the helicopter as offering help from above for the common man in trouble; or as a plaything of the rich that murders sleep; as an escape pad from a city under threat; or as a weapon that lures a feckless superpower into wars that it cannot win. After seven decades of operation, the "god machine" is the direction the helicopter has taken. It is not providing daily transport for the millions and it never will. What are the outer limits of its usefulness? What are the consequences that no one expected? Helicopters did more than speed the human journey: they changed the way we see the world and expanded the range of risks we take. Namely, Americans entering the wilderness have come to depend on choppers to pull them out of trouble. Instead of calling in volunteer helicopters on a catch-as-catch-can basis, as with Link Luckett's 1960 rescue effort on Mount McKinley, authorities at Denali National Park pay to keep a world-class pilot and a high-performance helicopter on strip alert for three months each year.

Talkeetna, Alaska, is a small town off the main highway from Anchorage to Fairbanks and has two airports extremely close to each other. One is a gravel airstrip just off the main street of shops and cafés, where during the short season any visitor can buy lattes, sandwiches, souvenirs, and flight-seeing packages. Homeowners living along the strip fly their personal airplanes from here.

A short walk away, across the railroad tracks, is the city airport used by all larger aircraft. Here, on a June morning in 2005, Jim Hood is speaking to three rescue volunteers about using his white Lama helicopter on Mount McKinley, called Denali by the Athabascan natives. The phrase *Denali Lama* is painted on the lower front windscreen. Hood is a rescue pilot on contract to the National Park Service and, after due consultation with the park headquarters, will fly to retrieve injured or ill climbers if conditions allow.

The purpose of his talk is to give the volunteers a refresher about safe helicopter operations before they confront an emergency in some howling, frigid wind. Up on the Hill, as mountaineers call McKinley, is not the place or time for a class about helicopter basics. On the day I visit, the weather in Talkeetna is a very pleasant sixty-five degrees, but at this moment and less than forty-five minutes away via Hood's helicopter, the weather on the slope at 17,000 feet is a fifty-mile-per-hour wind at minus ten degrees Fahrenheit. The wind is no doubt more savage a half mile higher at the summit but researchers haven't been able to keep their anemometers from being torn away, so no one knows how loudly the wind roars up there when the big storms come. A wind speed in excess of 200 miles per hour is a good guess.

Hood, dressed in jeans, reptile-skin boots, and a black leather jacket, stands with one foot propped up on the skid of the landing gear and talks laconically about visibility in the landing zone, the need for good radio communications, and the need to mind the tail rotor when approaching a helicopter from a slope. Volunteers must police the site ahead of time for anything that could get caught in the rotorwash and swirl up into the rotors. Once the helicopter is launched, they must be aware that its fuel will be at the bare minimum to make the save and get back to the fuel supply at 7,000 feet.

That means no mountaineers' packs are allowed on board. "We're walking a real fine line, right there at the limits," Hood says. "Our job is simply to get 'em off where nobody else can do it, then we meet the medics at Seven Camp." When the occasion calls for it, Hood can bring up a device informally known as the Body Snatcher. It's an electrically driven claw that can bring back a climber's body.

These men are experienced climbers and have been on the Hill before, but if Hood were trying to make the point to novices about what an emergency is like up there, it would be worth a few minutes to walk down the road to the Talkeetna Cemetery, across from the K-2 Flying Service and set in a clearing in an evergreen forest. Here is the Climbers' Memorial, a testament to what happens to people on Mount McKinley when it is crowded with climbers and the full force of the jet stream comes down. Carved from cedar, it shows two mountain climbers going up a totem pole. It was erected in September 1992, the year a monumental storm dumped five feet of snow at middle elevations and hit the higher camps with 110-mile-per-hour winds. The storm lasted for ten days and seven people died during that time. It was the worst run of fatalities since the first successful climb in 1913. Rangers credited the Lama with saving six climbers after the weather broke. In 1992 alone the Lama was used on five major rescue efforts. Total rescue costs for stranded climbers, two-thirds of whom were foreign climbers, were $431,000.

According to mountain climber George Rodway, writing in the journal *Wilderness and Environmental Medicine,* the toll was in part an unanticipated, delayed result of the rescue in May 1960. "It is probably not an exaggeration to suggest that the Day/Bading incident changed the attitudes of many mountaineers interested in climbing McKinley.... Not surprisingly, all of this had somewhat of an adverse effect on the behavior of those who came to ascend the peak. The caution displayed by the pioneers, not to mention the acceptance of the fact that one's destiny was alone in one's hands, became a thing of the past among many modern McKinley climbers."

Helicopter assistance is such a part of the wallpaper now that the Park Service doesn't attempt to recover its costs after a rescue, nor does it require climbers to post a bond or carry rescue insurance. The former would cost more in attorney fees than could be recovered, and the latter might legally obligate the helicopter crew to fly without regard to risk. Instead the park requires each climbing party to sit through a sobering orientation at the ranger station. By all accounts this dash of cold water has reduced the number of deaths as well as the number of helicopter rescue missions needed. So the crisis of dying climbers appears to have been tamed, at least until the arrival of another high-altitude storm comparable to the monster of 1992.

Sometimes helicopters make the news even when no rescue is attempted, as on September 11, 2001, when police department helicopters orbited near the rooftops of the World Trade Center towers but made no attempt to touch down or to lower officers with rescue equipment to open a way for office workers to reach the roof. The situation laid out like this: The South Tower, downwind from the North Tower fires, was completely shrouded in smoke from the North Tower, and fell just seventy-three minutes from the first attack. If a helicopter rooftop rescue was in any way possible, it could only have aided those in the North Tower. Inside the North Tower people were alive and trapped above the impact zone. One was software executive Peter Mardikian, who at the time of the attack had been seated in the restaurant Windows on the World for a breakfast conference, on the 106th floor of the North Tower. Mardikian called his wife from there and told her that he was going to try and get to the roof.

Telephone calls and electronic messages from the upper floors soon revealed that no one could get there. To get from the topmost accessible landing of the stairwell all the way to the roof would have required a person to get past three locked steel doors, or, alternatively, to pass through one steel door and then activate a large platform elevator used to raise the window-washing rig. Without a signal from the Security Command Center on the twenty-second floor, these steel doors wouldn't open, and when the command center made an attempt to do so at 9:30 A.M. the electronics were so damaged the signal did not reach the roof doors. Even the fire axes available to broadcast technicians working on the 110th floor couldn't have broken through the two doors separating them from the North Tower roof. That would have required a "rabbit tool," a hand-operated hydraulic device used by emergency workers to get past such doors.

Why were the doors locked and operated remotely in the first place? As a state agency, the Port Authority was exempt from the city's fire code, which otherwise insisted that any occupant be able to reach the roof on his or her own without having to break down a door. The Port Authority wanted this arrangement because it was concerned about suicidal people leaping from the roof, vandalism, thrill seekers, and politically motivated attacks on the huge array of broadcast equipment atop the North Tower. Politically motivated attacks on radio transmitters had happened before, such as in Peru in 1981 and in Lebanon in 1985. Further, this thinking was in line with the preferred "defend in place" strategy, which calls for high-rise occupants to stay where they are during a fire unless and until those on specific floors are instructed to head for the street. Though no helicopters would be available for succor on 9/11, ironically, the Port Authority's tourist-attraction contractor had set up a helicopter simulator ride in the visitors' gallery just below the roof of the South Tower.

Following the September 11 attacks, the fire department and Port Authority defended the locked doors by saying that police helicopters couldn't have accomplished anything anyway, because of smoke, heat, turbulence, and oxygen starvation that could cause jet engines to flame out. Police Commissioner Bernard Kerik seconded this notion, telling the press that pilots were under orders from their commanders not to approach the rooftops on their own.

There was a time when many firemen, including some members of the FDNY, expected helicopters to do much in case of fire in the vertical city. "Smokey Joe" Martin, a legend among New York firemen of the early twentieth century, once expressed the hope that helicopters would be useful in bringing hoses to upper stories at skyscraper fires. Today, only a few departments—Los Angeles is one—routinely train firefighters to land on skyscraper rooftops. New York is not one of those. Following the high-rise fires at São Paulo, Brazil, John T. O'Hagan of the FDNY was concerned that the lure of helicopter rescue could lead to bad decisions if it encouraged people in the upper stories of skyscrapers to go up the stairs when they should stay where they were, or, if instructed, go down the stairs. Such doubts persist today in New York.

Similarly, some observers of the Vietnam War grew concerned that helicopters eroded good leadership by tempting infantry officers to rise above the fray in their personal "command and control" helicopters.

Commanders who are comfortably distant from the battlefield have always been a concern to soldiers. The subject came up in the 1946 novel *Private Angelo,* by Eric Linklater. In the story, Angelo is a soldier in the Italian Army stationed in Calabria and must overcome his cowardice to fight the invading English Eighth Army and the Germans too. His commander is Colonel Piccologrando, who throughout the war lives in his villa in distant Rome, and issues tactical orders from there. The analogs in Vietnam were officers who lived in Saigon and monitored the war from the Circle Sportif tennis club in Saigon. But helicopters are equally capable of landing near a firefight when the commanders want a closer look. After hearing that a company commander was dead in battle and his battalion commander's helicopter pilot refused to land because the gunfire was too heavy, Colonel Hank "Gunfighter" Emerson landed and waded into the fight, using tracers from his M-16 to direct machine gunners where to place their fire. "Helicopter [command and control] is a great tool," Emerson says, "but you've got to land and show yourself to the troops."

As the commander of the Second Infantry Division in Korea five years later, Emerson summoned his troops to remind them that the North Koreans could cross the nearby Demilitarized Zone at any time, and if so he expected his officers to land their command and control helicopters from time to time and see the battle from the ground. "You should have heard the cheer from the troops," Emerson recalls. "This was after Vietnam so a lot of them knew the problem." Emerson has a name for that problem; he calls such commanders "airborne cheerleaders."

"I can see him now, with the veins popping out in his neck," retired General Colin Powell says. Powell was a battalion commander for Emerson at Camp Casey. "Gunfighter was like that, always leaning forward. He'd say, 'If the North Koreans come across, we're not going to sit on our side of the DMZ and take it.... We'll let the South Koreans be the line of resistance and meanwhile we'll go deep with helicopters and use air mobility to get behind them and cut them off.' That was the Gunfighter's angle: we're going to go deep. He was like a football coach, full of exciting ideas. But here we're the commanders and we never figured out what we were supposed to do next if that actually happened—how would we find our positions, and how would we get resupplied?

"Really, I think he was making a rhetorical point: 'Guys, it's important to be with your troops.' That was the Gunfighter all the way," Powell says. "But you can't control your battalion if you're stuck in the swamps. You have to know when to be on the ground and when to be looking over the battlefield."

Today's helicopters continue to be dogged by high volumes of low-tech enemy weaponry, at least when traveling near the ground and at low speed. Thirty-one of thirty-two Apache Longbow helicopters were damaged by heavy ground fire on March 24, 2003, near Najaf, Iraq, as the aircraft slowed to fire Hellfire missiles at the Second Armored Brigade of the Medina Division. One crash-landed after a bullet damaged its flight-management computer. Later that day Iraqi television showed a film of an elderly farmer of Najaf who, they said, had shot the Americans down with an old bolt-action rifle. While that particular claim was ridiculous, the damage from massed ground fire was very real. The air force, still stewing after all these years about the army's rotary-winged air-armada, did what it could to publicize the fiasco to the army's chagrin.

"We still expected the enemy to look, smell, taste, and move like the enemy," Eleventh Regiment commander William Wolf said after the battle. "We didn't expect hundreds, if not thousands, of small arms hiding behind trees, hiding behind farm walls, using lights to an advantage, hiding in the corners of a small village, shooting from the tops of schoolhouses."

Such gunmen may soon have fewer pilots at whom to shoot. The U.S. military is paying for the development of unmanned helicopters for high-risk combat missions, which will avoid exposing men to heavy gunfire during battlefield rescues and resupply. The long-term plan is to make these aircraft "autonomous," meaning that humans would give the machine only a mission outline and then stand back as the aircraft flies off. One such proposal is called the Fire Scout, which when fielded should be smaller than a two-seater personal helicopter. Already a prototype has flown itself to a landing on a moving aircraft carrier.

The idea of nimble, self-piloting helicopters that slip through gunfire to render aid on the battlefield has strong appeal, but meanwhile work is still needed on the traditional manned helicopter. Goals include lower fuel consumption, better crashworthiness, less aerodynamic drag, and less rotor noise. The latter might come from redesigned rotor tips, computer-controlled blades, or even the acoustic trick called "active" noise cancellation.

No cost-cutting breakthrough is in sight, but if found, it still wouldn't jump all the hurdles facing the home helicopter. When I first took an interest in helicopters I talked with my children about what it would be like to have one parked in the backyard. The kids agreed unanimously that it would be fun when traveling to their grandparents on holidays. At lunchtime along the way we could hover over a fast-food restaurant and send a boy down the rope ladder for a bag of burgers. Their mom could fly them on errands and impress the neighbors, in the manner that Chuck Marthens of Arizona Helicopter Service had flown his boss's daughter to elementary school in 1947. That early Bell 47 was a family flying flivver, a "touch of things to come," predicted *American Helicopter* magazine.

A helicopter pilot's manual and then a few hours flying with my instructor John Lancaster,* disabused me of both notions, however fun to contemplate. Flying a helicopter is not as carefree as in the movies. Before going anywhere into the wild blue my wife would be obligated to roll the family copter out of the backyard hangar, do a preflight inspection, warm up the engine for two minutes, engage the rotor, finish the checklist, do a hover check, and make a radio call. Only then, twenty minutes later, would she be at liberty to execute the mission. Back home she would have to sit in the helicopter another five minutes to cool off the engine before shutting it down (according to my instructor, who was solicitous of the engine's exhaust-valve seats). By any measure it would have been faster for her to walk them to school.

If brand-new pilots are planning to show off by dropping in on friends' backyards, they should think again. Wires are difficult to see from the air, many suburbs prohibit such stunts, and at any rate pilots are responsible for any damage or injury. Even experienced pilots in

---

* John H. Lancaster left Hummingbird Aviation in 2006 and took a job as a pilot flying a single-engine turbine helicopter between rigs in the Gulf of Mexico. He died in a crash on February 12, 2007, with one passenger.

built-up areas like Southern California and coastal New Jersey find it daunting to make any spontaneous, off-airport landings. Making helicopter owners and pilots most nervous are laws that go beyond limiting such stunts and block all new heliports. It's a problem for the industry because helicopters that are confined to airports have lost much of their edge over airplanes.

While of little solace to those living in major urban areas, helicopter pilots have considerable freedom to land in the vast unpeopled acreage lying between the coasts and outside of the cities, as long as they follow rules such as those restricting helicopter flights in wilderness areas and across certain parks.*

"I've landed at remote areas near restaurants in the Midwest," says Patrick Corr, president of the Helicopter Adventures private training school in Florida. "Generally their attitude is 'That's really cool, I wish I could do that.' But not on the East and West Coasts. Instead, it's 'Why did you land here? Did you have permission?'" Instructor Frank Barnes prefers construction sites during his cross-country flights because these usually have portable rest rooms.

"I don't do unplanned back-lot landings because I don't know the municipal attitude," says Harrison Ford. "They all have to have a say in helicopter landings. You're walking into trouble to do that kind of thing."

Over the last two decades pilot Ken Johnson has seen a sharp rise in helicopter opposition near Jackson, Wyoming. "Now there are a lot of what I call 'greenies.' I get sideways about that," he says. "I do a lot of rescue work in the mountains and these are the same kind of people that call for help. It's ironic. We had a two-year battle over maintaining the heliport, because the Conservation Alliance for the Greater Yellowstone Ecosystem fought it. The funny thing is, helicopters can't just do search and rescue. They have to do a whole lot of things to survive. If helicopters did just search and rescue, who would pay the costs?"

The counterargument is that publicly funded helicopters could handle search and rescue, but there have been cases where commercial

* Orlando Helicopter Airways modified a small number of medium transports to serve as flying RVs, which it dubbed the Heli-Camper. The larger of the models employed a Sikorsky S-58 airframe. It offered sleeping accommodations for up to six people, a small kitchen, bathroom with shower, air-conditioning, generator, entertainment center, and a screen door.

and private pilots responded first because they were close by and public helicopters weren't available in time. One example is the privately owned helicopter that responded to the Dupont Plaza skyscraper fire at the authorities' request.

Throughout his third career as helicopter pioneer, Igor Sikorsky collected clippings of helicopter rescues, and had his staff maintain a running tally of all lives saved by helicopter. It's approaching 2 million. On the day he died in October 1972, Sikorsky was drafting a congratulatory letter to pilots who had assisted at the Andraus Building fire. It was the last of thousands of such notes that he had dispatched over the years. "Were Igor Sikorsky here," Sergei Sikorsky said of his father in a 2005 speech, "I think he would say that his greatest source of satisfaction is the fact that the helicopter proved almost immediately to be a unique instrument for the saving of human lives."

What didn't delight Igor Sikorsky was the evolution of some helicopters into armed predators, and the common portrayal of helicopters in popular fiction as something menacing, or dangerous, or both.* "In movies, the helicopter is either flying in army dudes or it's blowing up," says Larry Welk.

Sikorsky's concept of the helicopter as the ultimate rescue machine was linked to his broader feeling that helicopters would in time liberate ordinary people from the tyranny of gravity and distance, allowing them to explore the hidden places of the world without the need to lay roads to get there. It explains his interest in the family camping helicopter, complete with balcony for sightseeing. Reaching remote places, and active volcanoes in particular, fascinated him. In the summer of 1945 Sikorsky joined a helicopter expedition to hover over the new crater atop Mexico's Paricutín volcano.

Sikorsky's conviction that helicopters would destroy space for people—what might be thought of as a liberation theology for the aerial age—has truly come to pass for some people, and not just the wealthy or powerful. Among the modern helicoptrians are Mike Kunz and his merry band of Arctic archaeologists, who gather on the vast northern slope of the Brooks Range each summer to scout for

---

* One early film using helicopters as performers was the made-for-television movie *Birds of Prey* (1973), in which one helicopter chases another.

promising sites of ancient origin on behalf of the Bureau of Land Management. Their turf is 200 miles from north to south and 250 miles from east to west.

In any given season, Kunz's crew uses its two leased helicopters to move whatever needs moving, much like a family uses a van or car. When each season opens, the first task is hauling gear for one or more field camps. While cargo aircraft from Fairbanks can bring supplies to a base camp, thanks to a long gravel strip originally built for an oil-exploration project, his summer field camps far to the west are temporary and offer no place for airplanes, even small ones, to land. Since a typical field camp is 140 miles west of the base camp and backpacking out there would take a hard two weeks of walking across mushy ground, one way, helicopters do the hauling.

The mundane nature of the helicopters' work would delight a flying-flivver enthusiast of the 1940s. When the field camp is going strong and the cook needs water, she loads the cabin of a helicopter with blue plastic jugs and sends the pilot off to fill them at the nearest river. When grizzly bears pester the camp, the helicopter serves as a flying noisemaker that usually drives the animals away. Each helicopter is a van, bringing the occasional load of VIPs from the Department of the Interior. After they leave and blissful silence returns, each morning the helicopter flies Kunz's fellow investigators Tony Baker and John Dube off to work. That allows the two men to walk a different route each day across the hills, ridges, and river flats. Baker and Dube map the promising sites for some future team to examine and perhaps excavate. So far, all the years of survey work that Kunz has supervised account for 5 percent of the territory available. That leaves 47,000 square miles of the National Petroleum Reserve–Alaska to go. The land is so vast that Dube and Baker and their successors could walk new ground every summer day for a hundred years and not finish the survey.

Already the years of walking and helicoptering have turned up remarkable things: most notably, evidence of the oldest known human activity in the Western Hemisphere. The discovery happened in 1978. "It was a cloudy, misty day like this one when I found Mesa," recalls Kunz as we are shoveling snow to clear off tent platforms at the base camp. He points out the site, about five miles away: a low, black hill. "I was doing a survey project on a plan to crush igneous rock to gravel for the top of the runway, so I was looking at resources that might be used for that. I was dropped in by helicopter and walked five miles that day. I reached Mesa, climbed up, and found a projectile point right off. It was one of the most important finds in the last forty to fifty years in North America."

In time Kunz's team uncovered dozens of shallow fire pits atop the steep-sided hill of lava-like rock. Hunters of the late Pleistocene took advantage of its clear field of view to watch for bison and caribou moving across the dry, grassy landscape. At 150 miles north of the Arctic Circle, the spot on the knoll at the time of discovery was so remote and eerily peaceful that nothing had disturbed the ash and stone points left in the campfire hearths across the span of 11,200 years.

I don't have a full day to walk to Mesa and back across the wet ground, which is dotted by thousands of camel-like humps called tussocks that can easily twist an ankle. I think: If I had a helicopter up here it would be a fine way to explore the niches of the Range. It would be like having the "seven-league boots" described in the seventeenth-century fairy tale "Tom Thumb" by Charles Perrault. The legendary boots would allow any wearer to take twenty-one miles at a single stride. In Perrault's story, the hero, Tom Thumb, is a small and wily child who steals the boots from an ogre, then uses them to make a fortune back home in expediting the delivery of information.

With a personal helicopter I could alight delicately on the Mesa and appreciate the kind of view across the river flats that Pleistocene hunters once had. I would take the doors off and feel the wind in my face like Igor on his high balcony. I would vault across Otuk Creek— now impassable due to deep, frigid meltwater from the snowpack— to inspect an exploratory drilling site last used twenty-five years ago, up a rocky slope in the Ivotuk Hills. While exploring far to the east I would be on the lookout for a rumored leftover from the DEW Line air-raid warning network. I've heard that one radar station was abandoned in place and everything lies unchanged from the fifties, as if emulating the timeless hearths left by those long-ago hunters atop the Mesa.

There is one helicopter hanging about the camp today, and the décor inside is showing its age like Perrault's seven-league boots, but it's much too busy for joyriding. Mike Kunz is loading the aircraft with a gasoline generator, cans of fuel, and rough wooden timbers. He straps the timbers to the skids with parachute cord, bound with a tidy knot that might be called a helicopter hitch. Pilot "Mad Mel" Campbell in his blaze-orange jumpsuit stubs out one last smoke, joins Kunz in the helicopter, and lights the turbine. In a few moments Campbell lifts the collective; though the ship has not yet lifted, I can tell this from the distinctive sound, changing from whine to whap, as the two blades first grip the air. The helicopter rises to a hover. Inside, Campbell scans the engine instruments. It's a go. Kunz on the left side gives a salute to us mortals on the ground, that humble place that early Italian balloonist Vincenzo Lunardi called "the lower world." The helicopter leans forward, accelerates quickly, and departs this isolated spot for one even more remote, far over the hills to the west.

Appendix I

Timeline of the Helicopter

21,000 B.C.E.: Weapons called "throwing sticks" (similar in appearance to boomerangs, but which do not return after being thrown) are used in what is now Poland.

50: Heron of Alexandria develops a simple rotating-tube device powered by steam called the aeolipile. It is the forerunner of the helicopter tipjet design, as in the Hiller Hornet.

1483: Leonardo da Vinci sketches a man-powered helicopter with a helical rotor, built around an iron frame and looking something like a carousel. Leonardo's sketch is misplaced and then rediscovered three centuries later.

1680: Dutch physicist Christian Huygens sketches an internal combustion engine relying on gunpowder for fuel.

1746: English engineer Benjamin Robins is the first to use a "whirling arm" to study the air resistance encountered by a moving object. He discovers that air friction is determined more by shape than by the frontal area of an object.

1754: Mikhail Lomonosov builds a spring-powered helicopter model that develops some lift while suspended at the end of a string. He demonstrates it to the Russian Academy of Sciences in hopes of building a full-sized model for lofting scientific instruments.

1765: French mathematician A. J. P. Paucton proposes using Archimedes' screw as a rotor for lift, anticipating the helicopter.

1784: François Bienvenu and a scientist named Launoy (possibly Claude Jean Veau de Launay) demonstrate a free-flying model helicopter in front of the French Academy of Sciences. It uses silk-covered rotor blades and is powered by a bow made of whalebone.

1791: John Barber designs a gas turbine with compressor. He plans to use it in a horseless carriage.

1794: First use of a balloon in wartime: *L'Entreprenant,* at Maubeuge, France.

1804: George Cayley flies the first fixed-wing, unmanned glider.

1809: George Cayley begins publishing a set of papers in *Nicholson's Journal* that frames the science of aerodynamics.

1828: Italian shoemaker Vittorio Sarti assembles a scale-model rotorcraft with two sets of sail-rotors, powered by jets of steam that emerge out of nozzles.

1842: Horatio Phillips flies a model helicopter that is powered by steam jetting from the blade tips.

1843: George Cayley publishes a design for the Aerial Carriage, a tandem helicopter, in *Mechanics' Magazine.*

1853: George Cayley's glider registers the first manned flight of a fixed-wing aircraft.

1855: William Rankine sets out the mathematics of ship propellers that provide lift or thrust. Later his principles will be adapted to the performance of helicopter rotors, called Momentum Theory.

1859: Edwin Drake drills the first successful oil well, in Pennsylvania.

1860: Belgian inventor Joseph Étienne Lenoir patents the first workable internal combustion engine.

1861: Mortimer Nelson of New York receives a patent for his "aerial car," a helicopter.

1862: Confederate soldier William Powers builds a scale-model mock-up of a war helicopter. He hopes a full-sized one will break the Union Navy blockade of Mobile Bay.

1862: Alphonse Beau de Rochas lays out the concept of the four-cycle gasoline engine, which will prove a radical improvement over Lenoir's design, and will be necessary for aircraft flight.

1863: The Heavier Than Air Society forms in Paris to finance and build a helicopter. Ponton D'Amecourt of the society is the first to create the name "hélicoptère."

1868: A sixteen-pound, one-horsepower steam engine built by F. J. Stringfellow is enough to keep a small airplane in flight at the Crystal Palace Exhibition.

1870: Alphonse Penaud builds an improved version of the Launoy-Bienvenu model helicopter by substituting a rubber-band motor for the whalebone bow. This toy goes on to entice the young Wright brothers into aviation.

1872: Francis Wenham presents a paper to the Aeronautical Society of Great Britain that points the way to the ideal helicopter blade.

1876: Nicolaus Otto invents a workable four-cycle gasoline engine.

1878: Enrico Forlanini flies a steam-piston-powered helicopter of a coaxial design, to a height of forty feet. It hovers for almost half a minute.

1884: The electrically powered airship *La France* makes the first round trip of any aircraft, out of Meudon, France.

1886: Jules Verne publishes *Robur the Conqueror*. It is the first book to describe rescue by helicopter.

1890: Elisio Del Valle patents a child's "clown toy" that rises into the air under rotors after being spun up by a string. Del Valle's toy is the first to employ hinged rotor blades, which later proves essential to autogiros and helicopters.

1891: Otto Lilienthal begins the first of 2,000 glider flights over his lifetime.

1896: Samuel Langley successfully demonstrates the unmanned, gasoline-powered airplane *Aerodrome No. 6,* which he launches from a houseboat in the Potomac. Later manned attempts do not succeed.

1901: Vast quantities of crude oil are discovered on the Gulf Coast of Texas, beginning with the Spindletop Field.

1904: Charles Renard (of the *La France* flight) proposes that the blades of rotor or propeller will work better if they are hinged to flap.

1906: The *Société Anonyme des Turbomoteurs* of France operates the first successful gas-turbine engine.

1906: Gaetano Crocco patents the idea of changing the pitch of helicopter rotor blades cyclically. Even without a workable helicopter to test his ideas, he forecasts that cyclic pitch will make it possible for helicopters to fly forward.

1907: Louis Bréguet and Charles Richet claim a successful hover of their *Gyroplane No. 1.* Bréguet discontinues his experiments in 1909, and will take the subject up once more two decades later.

1907: Paul Cornu claims his helicopter achieves a successful hover at a height of five feet, carrying pilot and later a passenger.

1909: Igor Sikorsky builds an unpiloted, coaxial helicopter in Kiev. It develops lift but does not get off the ground.

1909: Emile Berliner (inventor of the phonograph record) tests a wheeled helicopter-test rig on a track and discovers that rotor lift increases drastically when a helicopter moves forward. Later this is called "translational lift."

1909: Emile Berliner and John Newton Williams claim to have gotten their helicopter off the ground with one man aboard, while steadied.

1912: Boris Yur'yev assembles the first full-sized single-main-rotor helicopter, containing the essential elements of the swashplate mechanism for the hub of a helicopter rotor. Yur'yev is unable to bring it to flyable status.

1912: Jen Ellehammer builds a coaxial helicopter and makes verifiable short hovering flights.

1913: Ship expert Fred T. Jane writes that "except as a war-machine, the aeroplane is of little interest or use to anyone."

1917: An Austro-Hungarian team led by Stefan Petróczy claims to have flown a tethered helicopter for making battlefield observations, as an alternative to hydrogen-filled balloons.

1922: Raoul Pescara flies a helicopter inside a dirigible hangar at Issy les Moulineaux, France.

1922: The H-1 helicopter of George de Bothezat and Ivan Jerome flies in Ohio without assistance from ground handlers, and drifts downwind.

1923: Juan de la Cierva builds and flies the first successful autogiro, the C.4. He is the first to recognize that an unpowered main rotor can hold an aircraft in the air if the aircraft is pushed or pulled through the air by a separate propeller.

1924: Henry Berliner's No. 5 helicopter (augmented with fixed wings) hovers, and flies at speeds up to forty miles per hour.

1924: Louis Brennan flies a large two-bladed helicopter inside a building. It is powered by propellers located on the rotor blades.

1924: Étienne Oehmichen of France wins a major prize by flying his helicopter over a one-kilometer circuit. He averages five miles per hour.

1924: Following his successes in Russia, American immigrant Igor Sikorsky returns to building four-engine transport aircraft with the S-29.

1925: A. G. von Baumhauer of the Netherlands builds a single-main-rotor helicopter with a tail rotor to counteract torque and makes very brief flights. The two rotors use separate gasoline engines for power. While the helicopter is not a success, von Baumhauer incorporates many key elements of later helicopters: a swashplate in the rotor head to change rotor-blade pitch, hinged rotor blades, and cyclic control.

1925: Louis Brennan's helicopter crashes. Brennan and other helicopter inventors suspend further work, in part because Cierva's autogiro seems to hold more promise.

1928: Cierva flies from London to Paris, inaugurating the commercial promotion of autogiros.

1930: Unofficial flights of the Soviet TsAGI 1-EA helicopter begin. This helicopter relies on a single main rotor and smaller side-rotors for control.

1930: Corradino D'Ascanio flies a coaxial helicopter that relies on servo-flaps on the trailing edges of each rotor blade. This controls pitch of the blades by leverage.

1931: Former president Calvin Coolidge is irritated when noise from a newspaper's autogiro interrupts his speech at Warren G. Harding's tomb. It is the first rotorcraft noise complaint.

1931: Pitcairn Aircraft sells the first American-manufactured autogiro, the PCA-2.

1933: Belgian inventor Nicolas Florine flies a successful tandem-rotor helicopter.

1935: Juan de la Cierva produces an autogiro (the C.30) that can "jump" into the air using a main rotor that is powered only while the aircraft is on the ground.

1935: Austrian-born Raoul Hafner builds an autogiro (the AR-3) that demonstrates many vital control features of the helicopters to come, including cyclic and collective pitch built into a rotor hub with flapping blades. While the AR-3 was not capable of hovering, it was a major step forward.

1935: Louis Bréguet and Rene Dorand build the first helicopter generally accepted as meeting all the performance tests expected of a true helicopter, the "Gyroplane Laboratoire." Their coaxial design makes flights of over an hour, covering a distance of thirty miles.

1936: Henrich Focke's twin-rotor, side-by-side helicopter, the Fw 61, begins making test flights.

1936: Juan de la Cierva is killed in an airliner crash at Croydon Airport.

1937: The Fw 61 makes the first successful autorotation landing of a helicopter from cruising flight.

1938: Two German pilots fly the Fw 61 side-rotor helicopter inside the crowded Deutschlandhalle as part of an exposition.

1938: The U.S. Congress passes the Dorsey Bill to help fund the construction of helicopter prototypes.

1938: A Scottish helicopter, the Weir W-5, begins test flights. It is based on the Fw 61.

1939: Eastern Air Lines begins a one-year test with an autogiro that flies mail to and from the roof of the Philadelphia post office.

1940: Igor Sikorsky flies his first prototype helicopter. The cyclic controls do not work as intended and he substitutes with multiple tail rotors on outriggers.

1941: Igor Sikorsky is the first to fly a successful helicopter using a single main rotor and a single tail rotor to counter torque.

1942: Nazi Germany attempts to put helicopters into production, but repeated Allied bombing raids block all but a few of them from completion.

1942: After the British request a helicopter to assist in North Atlantic antisubmarine patrols, Sikorsky begins work on a larger helicopter, the R-5 Dragonfly.

1942: The Bell Model 30, *Ship 1,* begins test flights in Gardenville, New York.

1943: Frank Piasecki flies the single-seater PV-2 for a military audience at Washington National Airport, making the third successful American helicopter to fly. Piasecki receives a contract to develop a heavy-lift tandem helicopter for the navy (the PV-3 prototype).

1943: Newspaper supplements, speeches, and magazines carry a message that the family helicopter (price about $2,000) will be a postwar reality. During a speech to aeronautical engineers in New York, Igor Sikorsky forecasts sales of hundreds of thousands of helicopters after the war, as a new form of transportation rather than as a replacement to the car.

1944: First rescue via helicopter, using the R-4 Hoverfly. The location is a sandbar in Jamaica Bay, near New York.

1944: Stanley Hiller, age nineteen, begins test flights of his coaxial helicopter, the XH-44.

1944: A helicopter is first employed in war, in medical evacuations from Burma by the First Air Commando Group.

1945: The Sikorsky R-5 Dragonfly is delivered to the army but does not reach frontline units before the war ends.

1946: United Aircraft sponsors a commuting race in Connecticut matching helicopter, airline, train, and automobile. A Sikorsky helicopter wins.

1946: Senator H. Alexander Smith uses a helicopter briefly for campaign stops in northern New Jersey.

1947: First helicopter shuttle service to use a rooftop: Skyway Corporation in Boston, Massachusetts, which flies a Sikorsky S-51 between the Motor Mart Garage and Logan Airport. Neighbors of the Motor Mart file protests over the noise. The service shuts down four months later because of high operating costs compared to revenues.

1948: Lyndon Johnson startles small-town residents across Texas by visiting them in a helicopter during a hectic primary campaign.

1949: Successful trials of troop assault by helicopter, and the possibility of nuclear warfare, convince the Marine Corps to plan on using helicopter transports as an alternative to beach landings with amphibious craft.

1950: Hiller Helicopters begins testing the HJ-1, later known as the Hiller Hornet. Power comes from kerosene-fueled ramjets positioned at the tips of the rotor blades.

1950: First helicopter powered by gas turbines takes flight: Ariel III.

1950: Brain surgeon Dr. Valérie André begins flying a Hiller helicopter into Indochinese battlefields to evacuate the wounded.

1951: First use of helicopters to transport a large fighting force: twelve Sikorsky S-55 Chickasaws place 228 marines atop Hill 884 in North Korea.

1954: Test flights begin of the airplane-helicopter prototype 1-G (antecedent of the Osprey V-22). Builders are Bell Helicopter and Transcendental Aircraft Corporation.

1955: Sud Aviation begins test flights of the Alouette II, the first turbine-powered helicopter to enter production.

1956: First flight of prototype of Bell XH-40, which will become the UH-1 Huey.

1956: The federal government makes a permanent commitment to a network of high-speed interstate highways, with passage of the Federal-Aid Highway Act.

1957: The Fairey Rotodyne, a large convertible airplane-helicopter intended for passenger transport, begins test flights. It is powered by jet exhaust out the blade tips.

1958: First use of a helicopter as a platform for a television broadcast.

1958: An Alouette II firing Nord AS-11 guided missiles against insurgents in Algeria acts as the first all-up gunship.

1960: Alaskan pilot Link Luckett carries out a widely publicized rescue of two injured mountaineers, above 17,000 feet on Mount McKinley.

1961: USNS *Core* arrives in Vietnam with H-21 Flying Bananas.

1962: Turbine-powered helicopters begin arriving in Vietnam.

1963: Invading forces shoot down five American helicopters near the village of Ap Bac in South Vietnam, raising concerns that future air assaults may be more difficult than expected.

1963: First use of helicopters for rooftop rescue at a major high-rise fire (Hotel Roosevelt in Jacksonville, Florida).

1963: KTLA Telecopter broadcasts the collapse of the Baldwin Hills Dam in California, opening the era of live-in-the-sky breaking news.

1965: Battle of Ia Drang demonstrates the Sky Cavalry concept.

1966: Lakewood, California, begins the anticrime helicopter patrol called Sky Knight.

1967: Helicopter with a Santa Claus on board crashes at an Evansville, Indiana, shopping mall.

1968: Ninth Infantry Division uses Jitterbug airmobile tactics successfully against Vietcong main force units in the Mekong Delta.

1969: Israeli commandos in two CH-53 helicopter transports land at a radar installation at Ras Ghareb, Egypt, dismantle it, and bring the equipment back to Israel for study.

1969: Mil's Mi-12 prototype transport, the biggest ever flown, lifts forty-four tons.

1969: The Maryland State Police begins providing helicopter-ambulance services.

1973: Soviets test-fly the Mi-24 Hind. It begins with the dual role of troop transport and gunship.

1975: Helicopters prove essential to the evacuation of American civilians and guards from enclaves around Saigon, as the North Vietnamese finish their invasion of the South.

1975: American helicopters attempting to rescue hostages meet heavy antiaircraft fire from Khmer Rouge troops at Koh Tang Island, off the coast of Cambodia.

1978: Successful deployment of helicopter gunships and transports by Soviet forces during the Ogaden War persuades tacticians in that country that helicopters could turn the tide in future counterinsurgency actions.

1979: The Robinson R-22 personal helicopter receives an airworthiness certificate, authorizing inventor Frank Robinson to begin making and selling the low-priced two-seater.

1979: The H-60 Black Hawk, an all-purpose helicopter to supplant the Bell UH-1, enters service with the army.

1980: Operation Eagle Claw, intended to free U.S. hostages from imprisonment in Iran, meets with spectacular failure at the Desert One staging area when one helicopter collides with a C-130 transport.

1983: The Soviet Mi-26, the largest production helicopter in the world, enters service.

1983: Bell initiates the Osprey V-22 program, aimed at a "tilt-rotor" airplane-helicopter.

1984: High-altitude war relying on helicopters begins at the Siachen Glacier, between India and Pakistan.

1986: Mujahadeen forces begin using CIA-provided Stinger missiles against Soviet helicopters in Afghanistan.

1986: Dozens of Soviet transport helicopters drop sand and boron to help quell a runaway reaction at the Chernobyl plant, following a reactor explosion.

1988: The National Transportation Safety Board issues a special report calling attention to high crash rates among some air-ambulance operators.

1999: U.S. Coast Guard begins placing snipers on helicopters to stop drug-running boats in the Caribbean.

Appendix II
Cast of Notable Helicopters

Aérospatiale Lama: A French single-rotor helicopter that is popular for mountain work; it is a derivative of the early turbine-powered Alouette II.

Bell Model 47: Based on a two-bladed prototype called the Model 30, it was the first U.S. helicopter approved for sale. Its durability and stability in hover opened up the small utility helicopter market.

Bell UH-1 Huey: The iconic, all-purpose helicopter of the Vietnam War. First appeared there in 1962.

Bell JetRanger: A widely used single-turbine helicopter based on the Vietnam-era OH-58.

Berliner No. 5: Prototype side-rotor helicopter built by Henry Berliner that hopped and hovered during tests for navy officials at College Park, Maryland, in 1922.

Boeing CH-46 Sea Knight and CH-47 Chinook: Large, tandem-rotor transport helicopters powered by gas turbines.

Bréguet-Dorand Gyroplane Laboratoire: The first controllable, flyable helicopter that was verified in witnessed tests. However, it suffered damage during autorotation tests and the approach of World War II made continued repair and development impossible.

Cierva C.4: The first autogiro to fly successfully and an important branch on the helicopter's family tree. It led to a small number of improved, production autogiros that were sold for daily use.

De Bothezat H-1: a four-rotor helicopter financed by the U.S. Army that hovered successfully (and drifted sideways) in 1922 tests.

Eurocopter AS 350 B2: Known in Europe as the Ecureuil and in the U.S. as the AStar, it is a turbine-powered single-rotor helicopter commonly used for news reporting and air-taxi service.

Fairey Rotodyne: Innovative British "compound helicopter" of the late 1950s that relied on a helicopter mode for takeoff and landings (using rotor-mounted tipjets for power), then shifted to autogiro mode for cruise flight.

Focke-Wulf Fw 61: The first helicopter usable for travel. Only two of this side-rotor aircraft were built but the model became famous due to 1938 newsreels. It led to the Fa 223 Drache, the first helicopter useful for hauling cargo inside and outside the cabin.

Hafner AR-3: An evolutionary improvement on Juan de la Cierva's work. Raoul Hafner's autogiro of 1935 had a sophisticated rotor hub that anticipated what helicopters would need to fly.

Hiller Hornet: Prototype "tipjet helicopter," where the single main rotor was powered by ramjets mounted on the rotor-blade tips. It was simple in execution, but was noisy and used too much fuel given its small payload.

Hiller 360: Stanley Hiller's first production helicopter. Forerunner of the Hiller 12E single-rotor utility helicopter.

Kaman HH-43B Huskie: Turbine-powered, intermeshing-rotors helicopter used for rescues in the Vietnam War era.

Launoy-Bienvenu Machine Mécanique: a bow-powered, twin-rotor flying model demonstrated in 1784.

MD 500: Commercial version of the turbine-powered Hughes OH-6 Loach light observation helicopter.

Mil Mi-24 Hind: Soviet gunship used during the Soviet incursion into Afghanistan, often in coordination with Mi-8 and Mi-17 troop transports.

Mil Mi-26 Halo: Largest production helicopter in the world. Single-main-rotor type.

Pescara prototypes: Coaxial helicopter with four rotors on its mast, with each rotor having five blades. This prototype flew almost a half mile in April 1924.

Piasecki PV-3 Dogship: Prototype of the first tandem-rotor cargo helicopter. Eventually Piasecki's work led to today's Boeing tandem cargo helicopters.

Robinson R-22: Two-seat piston-powered helicopter widely used for personal travel and light-commercial applications including news-gathering, cattle-herding, and police patrol. A top-selling helicopter when grouped with the larger R-44.

Schweizer 300: Class of two-seat piston-engine helicopters often used for training. Originally based on the TH-55 Osage trainer.

Sikorsky VS-300: The first functional helicopter in America and the first to employ successfully a single main rotor and single tail rotor.

Sikorsky "R" models: The R-4 Hoverfly was a small helicopter that hauled supplies and made one-man rescues in the Pacific Theater. It was the first production helicopter ever used in a combat zone. The R-6 was an upgraded version of the R-4 that became available in the final months of World War II, and only in small numbers. The Sikorsky R-5 Dragonfly, which was three times heavier and had a glasshouse-style cabin, was designed for submarine hunting but did not contribute to the war effort. It was the basis for the civilian model S-51, which Lyndon B. Johnson used while campaigning for the Senate in 1948. The R-5 was used for rescues in Korea.

Sikorsky S-55 Chickasaw: A medium-capacity, loaf-shaped, piston-powered helicopter used heavily by the U.S. Marines in Korea. It was also called the H-19.

Sikorsky S-58: A more powerful piston-engine transport helicopter styled similarly to the S-55, and used in Vietnam. Later converted to turbine power and renamed the S-58T.

Sikorsky S-61: Civilian twin-engine, single-rotor helicopter used by helicopter airlines of the 1970s.

Sud Aviation Djinn: The only tip-powered helicopter ever to reach production.

TsAGI 1-EA: May have been one of the earliest flyable helicopters, but any flights of this Soviet prototype (circa 1930) did not occur in front of record-keeping officials.

Weir W-5: Scottish-built side-rotor helicopter of 1938, modeled on the Fw 61.

Notes

Introduction

1. The great majority of Aeschylus' plays are lost to history.
2. This helicopter, along with the Mi-6 Hook, was instrumental in dumping boron and sand to quell radiation after an explosion at the V. I. Lenin Chernobyl Nuclear Power Station at Pripyat, Ukraine.

Chapter One

1. Helicopter Adventures at Titusville, Florida.
2. By comparison, the Aérospatiale Lama, a jet turbine–powered helicopter suited for mountain work, burns sixty gallons per hour.

Chapter Two

1. While an instructor at St. Petersburg Academy, Lomonosov's protest against social customs there got him imprisoned. He wrote poems to the Empress Elizabeth until released in 1744. Lomonosov went on to write important papers—on heat, electricity, and the atmosphere—but the verdict of historians is that he blunted his impact with intemperate words and excessively broad interests.
2. Professors Michael Lynn and Patrice Bret introduced me to the scholarship in this area.
3. Whirling arms are the invention of mathematician and military engi-

neer Benjamin Robins, born in 1707. Robins used a modification of the whirling arm to measure the energy of balls fired from a cannon, and was the first to figure out that air resistance grows greatly as an object enters the transonic region, meaning that its speed approaches that of sound waves.

4. The Aerial Carriage would have been a "converti-plane" because the lifting rotors would have flattened into round wings by changing their blade pitch once in the air.
5. Patent No. 32,378.
6. Sometimes titled *Clipper of the Clouds*.

Chapter Three

1. In 1900 the attorney general for the state of Nebraska charged that the Standard Oil trust had been running a monopoly for at least eight years and was charging the state's consumers much more than it charged export customers, where it faced real competition.

Chapter Four

1. That show was the first major public exhibition for airplane makers.
2. Pronounced "breh-GAY."
3. Manufactured by Adams Company, Dubuque, Iowa. It is likely that Berliner's motor was the first radial engine built specifically for use on aircraft. The French-built rotary engine called the Gnome went on to power many Allied World War I airplanes.
4. Williams's vehicles had one notable feature: The driving engine, a rotary, was encased inside the rear wheel.
5. The commander of the balloon, Jean Marie Joseph Coutelle, cried out with delight about his view over the Austrian lines, but a few days later the Austrians set up the world's first antiballoon gun battery and forced him to lower his profile.
6. It was the same purse that had motivated Paul Cornu to build a helicopter a year earlier.
7. Airplane evacuations of French wounded began during the retreat from Serbia in 1915, but the use of air ambulances never became widespread during the war. A shortage of landing fields near the front was but one of the problems.

NOTES   293

Chapter Five

1. Today pilots call such behavior the "Watch-This Syndrome."
2. The rotors on American single-main-rotor helicopters turn counter-clockwise.
3. Cierva's family was irrevocably associated with Spain's Monarchist faction and the Monarchists went into decline after King Alfonso lost his throne in 1931.
4. Not a bestseller, this book was available only to licensees of his technology.
5. Pitcairn also invented the Mailwing, a small airplane ideal for moving mail to and from small airfields.
6. Other rooftop operations confronted similar problems; some added wind-directing vanes along the roof edge to reduce such turbulence.
7. D'Ascanio reached his peak of fame following the war as designer of the Vespa motor scooter, a product of postwar Italy that initially relied on surplus generator motors originally manufactured for fighter planes.
8. One example of the transferability of his ideas was the rotor hub of the Fw 61 helicopter, based on a Cierva design.

Chapter Six

1. The name was not an abbreviation of "helicopter" but rather a reference to the Swahili language.
2. Olga had persuaded her husband (a successful banker) to advance money for the project.
3. The initials stood for Vought-Sikorsky.

Chapter Seven

1. Nothing except a one-man helicopter like the Rotorcycle could have fit into such a space.
2. One demonstration of the fact that the All-American system could not fill the shoes of helicopters was the September 1943 recovery of twenty men from the Naga Hills of Burma after a C-46 transport suffered engine trouble and crashed while flying the Hump. One of those who parachuted out was CBS correspondent Eric Sevareid. The men and their rescuers ended up hiking out and there was no attempt to use the man harness.

3. Young's demonstration of hover control was indeed striking, but Étienne Oehmichen had patented the basic concept years earlier.
4. The HSL suffered from vibration, noise, and lack of usefulness in cramped shipboard settings. The navy declined to build its antisubmarine warfare program around it.
5. Pronounced "PIE-secki."
6. In 1943 the word *flivver* referred to a simple, no-frills automobile. Before the automotive connotation, the word referred to things breaking up or flying apart, as in "the steam engine came a-flivver."
7. By that time Piasecki's company had changed hands and was known as Boeing-Vertol. Boeing would go on to produce the big tandem, turbine-powered transports known as the CH-46 Sea Knight and CH-47 Chinook.
8. Pronounced "kah-MAN."

Chapter Eight

1. In that musical comedy Lahr portrayed a hapless airplane mechanic who sets a world altitude record in an "aerocopter," without intending to do so.
2. Great Lakes Greyhound Skyways, Inc.
3. Normally such a throng gathered only on Saturday evenings, when farming families flocked to town.
4. After Enstrom crashed his fifth prototype in 1958, Mrs. Enstrom gathered the fragments of the wooden rotor blades for firewood.
5. Cessna's helicopter was the CH-1 Skyhook, based on a design by Charles Seibel of Wichita, Kansas.

Chapter Nine

1. Estimates vary, but the total used in French Indochina was less than fifty. By contrast, the United States employed 12,000 helicopters during the Vietnam War. Another reason for the dearth was that American suppliers were busy filling orders for the war in Korea.
2. Israeli troops used helicopters in a similar lightning maneuver a year later, to block the retreat of Egyptian armor through Mitla Pass during the Sinai campaign.
3. Fuel consumption of tipjet helicopters is unavoidably twice that of ordinary helicopters of the same power, according to Professor Gordon Leishman of the University of Maryland.

4. NASA declined to pursue this idea, and opted for the semi-reusable space shuttle instead.
5. Still another benefit of gas turbines is less wear on the helicopter transmission, because the power arrives as a smooth torque rather than as pulses from a reciprocating mechanism.

Chapter Ten

1. One of the first men in Vietnam to begin collecting such songs was General Edward Lansdale, who recorded hundreds of songs as performed in his Saigon villa beginning in 1965. Beginning with the Huk Rebellion in the Philippines, the unconventional Lansdale won the reputation as a standout counterinsurgency operative because of his close study of the enemy's way of life and culture.
2. "The ARVN were underrated," according to anthropologist Gerald Hickey, who spent much time with them and understood the Vietnamese well. "Sometimes they went for months without being paid."
3. This was the Tonkin Gulf Resolution, a statement passed by Congress under the mistaken belief that North Vietnamese boats had attacked a U.S. warship.
4. At the time, Hickey was working for the RAND Corporation, which had a contract to advise MACV how to employ special forces troops to enlist the Montagnards in the war effort.
5. Second Battalion, First Brigade, 502nd Regiment, 101st Airborne Division. His commander was Major General Stanley "Swede" Larsen.
6. Taylor's father was General Maxwell Taylor, previously the American ambassador to Vietnam.
7. Emerson explains that a typical tactical situation could be pictured as broad stripes running north to south down a map. These indicated the likely routes of NVA troops.
8. These were arrayed in L-shaped patterns, and detonated by wire. Employing them required the ambushing squads to allow the enemy's point men to walk through the U.S. lines.
9. Kazickas later took a job with the Associated Press.
10. In one intelligence-gathering incident, an informer led Emerson's men to a building at which Vietcong commanders were meeting with a North Vietnamese officer. The men broke down the door, shot most of the enemy, and took the NVA officer prisoner.
11. Colonel Josiah Wallace and Lieutenant Colonel Robert Dirmeyer.

Chapter Eleven

1. Capital of the Nam Dong Province, northwest of Da Nang.
2. False radio calls, even NVA soldiers dressed in American flight suits, were common methods to lure helicopters into traps.
3. These were called Lima sites, and positioned in Laos as close as possible to the North Vietnamese border. Limas extended the Jolly Greens' working range greatly.
4. Along with some captured U.S. tanks.

Chapter Twelve

1. At fifty miles long and two miles wide, the Siachen is one of the world's largest glaciers outside of the polar regions.
2. Among the first was the California Ambulance Service in Los Angeles.
3. A representative of the fire department voiced those concerns in January 1979, during a cooperative project with the police department to survey and photograph twenty-five high-rise buildings that might be suited for rooftop helicopter landing zones.
4. In 1999, pilot Boyd Clines plucked an operator from a construction crane directly over a blazing five-story building in Atlanta. The radiant heat was so intense it softened the plastic chin bubble on his helicopter.
5. Including a flameout of the gas turbine. Peter Gillies warns his trainees that taking a helicopter through the plume of a large fire or smokestack exhaust has on multiple occasions starved the turbine engines of oxygen.
6. The Boeing-Vertol Model 107.
7. The early hope of helicoptrians, stated first by Manferd Burleigh of Greyhound in 1943 and repeated often in the business press for ten years following, had been that commercial helicopters would fly passengers at a cost of five cents per seat-mile. Actual costs far exceeded that figure.

Chapter Thirteen

1. During an economic crisis and associated crime wave in the 1990s, wealthy residents moved into guarded enclaves, fearful of a high murder rate and an even higher kidnapping rate. The largest guarded enclave, Alphaville, is walled and patrolled by over a thousand security guards.

2. As a current government-industry research plan says, "reduction of external noise generated by rotorcraft is rapidly becoming a major market driver worldwide as community tolerance to noise decreases."

3. The apartment building originally was constructed to be affordable, quality housing for low-income New Yorkers. Philanthropist-investors built it at a loss.

4. Proud inventor of the chess move called the Rice Gambit. Rice had contractors build a soundproof room in the basement of their house, relaying his moves to British opponents by telegraph. He sponsored tournaments on the condition that the contenders use the Rice Gambit.

5. Grant's OH-6 was surplused to the San Bernadino County Sheriff's Department, resold, upgraded, recertified for sale. Military helicopters, when surplused to local governments, must stay in "restricted class" until upgraded for sale on the market.

Chapter Fourteen

1. Organizers of such counterterrorism training missions commonly inform local police in advance but then ask them not to tell the public, out of concern that curiosity-seekers may interfere with the training or put themselves at risk.

2. North Memorial Medical Center in Minneapolis.

3. The CIA had two of them assembled for the mission. The official story is that the silencing modifications were removed and the two helicopters were sold into civilian service.

4. The prehistory includes the use of autogiros by newsreels and newspapers. After a saboteur set a fire on board the *Morro Castle* in 1934, near the New Jersey coast, an autogiro arrived to photograph it for the newspapers.

5. An example of helicopter-on-helicopter news is footage relayed in 1999 to CNN by two news-copters. These photographed a third helicopter rescuing a construction worker stranded on a crane during a building fire in Atlanta.

6. One occupant removed from the roof was a building engineer who informed the fire department that a set of fire pumps in the basement needed to be turned on manually. That information was important to saving the building.

Chapter Fifteen

1. In the thin air, this required a speed over the ground of about thirty miles per hour to maintain altitude.
2. Wayne Brown, now assistant chief instructor at Bell Helicopter's Training Academy, estimates that he has executed the "full-down" autorotation maneuver 80,000 times. He is one of Harrison Ford's refresher-training instructors.
3. According to Ballou, he brought along a correspondent to see the work of Aero Scouts and the flight made the correspondent violently ill. Ballou says he flew no more reporters after that.
4. A typical task required setting steel girders with high precision.
5. McElhinney was an LAPD pilot on duty during the Watts riots, tracking crowd behavior.

Bibliography

Introduction

"13 Die as Oil-Rig Rescue Fails in Gulf," *New York Times,* April 17, 1976, 1.

"Abandon Rig! The Loss of the *Ocean Express,*" *Proceedings of the Marine Safety Council,* November 1978, 115.

Booth, Tony. "Operation Chernobyl," *Flight International,* July 31, 1996, 40.

Campbell, Mel. 2005. Interviews by author. June.

Chondros, Thomas G. 2006. Personal communication. 8 February.

————. "'Deus-Ex-Machina' Reconstruction and Dynamics," in *Proceedings, International Symposium on History of Machines and Mechanisms,* 2004, 87.

Cubanski, Edward J. "Coast Guard HITRON—A Model of Success," *Sea Power,* August 2002, 39.

Culver, John C., and John Hyde. *American Dreamer: The Life and Times of Henry A. Wallace.* New York: W. W. Norton, 2000.

Davidson, D. S. "Australian Throwing-Sticks, Throwing-Clubs, and Boomerangs." *American Anthropologist,* January–March 1936, 76.

————. "Is the Boomerang Oriental?" *Journal of the American Oriental Society,* June 1935, 163.

Fales, E. D. "The *Ocean Express* Disaster—A Hard Lesson at Sea," *Popular Mechanics,* December 1980, 86.

Green, Stanley. *Encyclopedia of the Musical Theater.* New York: Da Capo Press, 1980.

Halloran, Richard. "Odds Said to Favor Cocaine Smugglers," *New York Times,* December 8, 1988, A23.

Hancock, Richard. 2005. Interview by author. 9 August.

Hessler, Peter. "The Nomad Vote," *The New Yorker,* July 16, 2001, 59.

Kieran, John. "A Boom in Boomerangs," *New York Times,* January 22, 1941, 27.

"Long Island Heroes: The 106th Air Rescue Group," *Newsday,* December 14, 1994, A32.

Lovering, Joseph. "On the Australian Weapon Called the Boomerang," *American Almanac and Repository of Useful Knowledge,* 1859, 67.

MacLeod, Steve. "Survivor from Sunken Ukrainian Ship Delivered to Shore: Hopes Dim for 30 Sailors Still Missing," *Ottawa Citizen,* December 11, 1994, A5.

Mastronarde, Donald J. "Actors on High: The Skene Roof, the Crane, and the Gods in Attic Drama." *Classical Antiquity,* October 1990, 247.

Mitchell, Robin. "Officials Drawing a Bead on Drug Runs," *St. Petersburg Times,* September 14, 1999, 1B.

Moffett, Cleveland. "Louis Brennan's Mono-Rail Car," *McClure's Magazine,* December 1907, 163.

National Transportation Safety Board. *Capsizing and Sinking of the Self-Elevating Mobile Offshore Drilling Unit Ocean Express Near Port O'Connor,* Texas, April 15, 1976. Washington, D.C.: NTSB, 1979.

"Officer's Wife Flies with Wilbur Wright," *New York Times,* October 28, 1909, 4.

Pociask, Martin J. "HAI Members Respond to the Fury of Katrina," *Rotor,* Winter 2005–2006, 16.

"Presenting . . . Raoul Hafner," *American Helicopter,* August 1947, 19.

"Promised Wonders of the Gyroscope on Land and Sea," *Current Literature,* July 1907, 90.

Purpura, Paul. "National Guard Flies to Rescue After Hurricane," *Times-Picayune,* January 8, 2006, 22.

Robinson, Linda. "The Coast Guard's New Secret Weapon," *U.S. News & World Report,* March 20, 2000, 38.

Robkin, A. L. H. "That Magnificent Flying Machine: On the Nature of the Mechane of the Theatre of Dionysos at Athens," *Archaeology News,* 1979, 1.

Rogers, Kathryn. "Copter Crash Kills 2 Trying to Bring Aid," *St. Louis Post-Dispatch,* August 10, 1992, 1A.

Shields, Rachael. 2005. Interview by author. 11 June.

Sterba, James P. "Coast Guard Moves to Investigate Sinking of Oil Rig," *New York Times,* April 18, 1976, 29.

Stout, David. "Coast Guard Using Sharpshooters to Stop Boats," *New York Times,* September 14, 1999, A18.

Subrahmaniam, Vidya. "It Is a Sky-High Campaign Here," *The Hindu*, February 18, 2005, 1.

"The Railway of the Future," *Outlook*, November 27, 1909, 641.

Thurston, Harry. *The World of the Hummingbird*. Vancouver: Greystone Books, 1999.

"Upper Paleolithic Boomerang Made of a Mammoth Tusk in South Poland," *Nature*, Vol. 329, No. 6138, 1987, 436.

Weiner, Eric. "Airborne Drug War Is at a Stalemate," *New York Times*, July 30, 1989, 1.

Weissman, Richard. "New Fight on Drug Traffic," *New York Times*, January 6, 1985, L1.

"World's Oldest Boomerang Found in Poland," *San Francisco Chronicle*, October 1, 1987, A40.

Chapter One

Casper, Willie. 2006. Personal communication. 9 June.

Coleman, Mike. 2005. Interview by author. 7 February.

Connor, Roger. 2005 and 2006. Interviews and personal communication with author. June–July.

Coyle, Shawn. 2006. Interview by author. 15 March.

Coyle, Shawn. *Cyclic & Collective: More Art and Science of Flying Helicopters*. Mojave, CA: Helobooks, 2004.

Johnson, Wayne. *Helicopter Theory*. Princeton, NJ: Princeton University Press, 1980.

Lancaster, John. 2005. Interviews by author. July–September.

Leishman, Gordon. 2005. Interview and personal communication with author. October–January.

Newman, Simon. *The Foundations of Helicopter Flight*. New York: Halsted Press, 1994.

"Pilot Recognized for Rescue at the Helm of a Malfunctioning Helicopter," *Wisconsin State Journal*, November 16, 2000, B4.

Prouty, Ray. 2005. Personal communications. 3 December.

———. *Helicopter Aerodynamics*. Cincinnati, OH: PJS Publications, 1985.

Prouty, Ray, and H. Curtiss. "Helicopter Control Systems: A History," *Journal of Guidance Control Dynamics*, 2003, 12.

Spenser, Jay. 2006. Interview by author. 10 June.

Tweedt, Barbara. 2006. Personal communication. 7 April.

Whyte, Greg. *Fatal Traps for Helicopter Pilots*. New York: McGraw-Hill, 2006.

Chapter Two

Ackroyd, J. A. D. "Sir George Cayley, the Father of Aeronautics. Part 2. Cayley's Aeroplanes," *Notes and Records of the Royal Society of London,* September 2002, 333.

Boulet, Jean. *The History of the Helicopter as Told by Its Pioneers 1907–1956.* Paris: Editions France-Empire, 1984.

Bret, Patrice. "Un Bateleur de la Science: Le 'Machiniste-Physicien' François Bienvenu et la Diffusion de Franklin et Lavoisier," *Annales Historiques de la Révolution Française,* 2004, 95.

————. 2006. Personal communication. 7 August.

Cayley, George. *Aeronautical and Miscellaneous Note-book of Sir George Cayley.* Cambridge, UK: The Newcomen Society, 1933.

————. "On Aerial Navigation," *Nicholson's Journal of Natural Philosophy, Chemistry and the Arts,* November 1809, 164.

————. "On Aerial Navigation," *Nicholson's Journal of Natural Philosophy, Chemistry and the Arts,* February 1810, 81.

————. "On Aerial Navigation," *Nicholson's Journal of Natural Philosophy, Chemistry and the Arts,* March 1810, 161.

Chambriard, Pascal. "L'Embouteillage des Eaux Minerales: Quatre Siècles d'Histoire," *Annales des Mines,* May 1998, 20.

Dumas, Maurice, *Scientific Instruments of the Seventeenth and Eighteenth Centuries.* New York: Praeger Publishers, 1972.

Emery, Clark. "The Background of Tennyson's 'Airy Navies.'" *Isis,* 1966, 139.

Gibbs-Smith, Charles H. "Sir George Cayley: 'Father of Aerial Navigation' (1773–1857)," *Notes and Records of the Royal Society of London,* May 1962, 36.

Greffe, Florence. 2006. Interviews by author. June–July.

Hahn, Roger. 2006. Personal communication. 20 July.

Howard, Michael. *The Franco-Prussian War.* London: Rupert Hart-Davis, 1961.

Instruction sur la nouvelle Machine inventée par MM. Launoy, Naturaliste, & Bienvenu, Machiniste-Physicien. Paris: Francois Bienvenu, 1784.

Kelly, Fred C. (ed). *Miracle at Kitty Hawk.* New York: Da Capo Press, 1996.

Lambermont, Paul, and Anthony Pirie. *Helicopters and Autogiros of the World.* London: Cassel, 1958.

Leishman, J. Gordon. *Principles of Helicopter Aerodynamics.* New York: Cambridge University Press, 2000.

Liberatore, E. K. *Helicopters Before Helicopters.* Malabar, FL: Krieger Publishing Co., 1998.

Liptrot, R. N. "Historical Development of Helicopters," *American Helicopter,* March 1947, 12.

Lopez, Donald, and Walter J. Boyne. *Vertical Flight: The Age of the Helicopter.* Washington, D.C.: Smithsonian Institution Press, 1984.

Lynn, Michael R. "Divining the Enlightenment: Public Opinion and Popular Science in Old Regime France." *Isis,* March 2001, 34.

———. "Public Lecture Courses in Enlightenment France." *The Historian,* Winter 2002, 335.

———. 2006. Interview by author and personal communications. June–July.

McClellan, James. 2006. Personal communication. 2 September.

"Mikhail Vasilevich Lomonosov," *Encyclopedia of World Biography,* Vol. 9. Detroit: Gale Research, 1998.

Munson, Kenneth. *Helicopters and Other Rotorcraft Since 1907.* London: Blandford Press, 1968.

Nelson, Mortimer. *Mortimer Nelson's Aerial Car.* New York: Mortimer Nelson, 1860.

Regourd, François. 2006. Personal communication. 8 September.

"Sir George Cayley's Aerial Carriage," *Mechanics' Magazine,* April 8, 1843, 36.

Tassin, Christian. 2006. Personal communication. 3 August.

"The Inflated Giant," *The Albion,* December 12, 1863, 598.

Verne, Jules. *Robur the Conqueror; or, A Trip Round the World in a Flying Machine.* New York: G. Munro, 1887.

Wright, Orville. "How We Invented the Airplane," *Harper's,* June 1953, 25.

Chapter Three

"AP-4103 Model Research, Volume 1," at http://history.nasa.gov/SP-4103/ch4.htm (Accessed July 19, 2005).

Chiles, James R. "Spindletop," *American Heritage of Invention and Technology,* Summer 1989, 34.

Christopher, John. *Balloons at War.* Gloucestershire, UK: Tempus Publishing, 2005.

Corn, Joseph J. *The Winged Gospel: America's Romance with Aviation, 1900–1950.* New York: Oxford University Press, 1983.

Crouch, Tom D. *The Eagle Aloft.* Washington, D.C.: Smithsonian Institution Press, 1983.

"Flying Machines in the Future," *Scientific American,* September 8, 1860, 165.

Gablehouse, Charles. *Helicopters and Autogiros: A Chronicle of Rotating Wing Aircraft.* New York: J. B. Lippincott, 1967.

Gillespie, Richard. "Ballooning in France and Britain," *Isis,* June 1984, 248.

Hallion, Richard P. *Taking Flight: Inventing the Aerial Age, from Antiquity Through the First World War.* New York: Oxford University Press, 2003.

Hunt, T. Sterry. "On the History of Petroleum or Rock Oil," *American Journal of Pharmacy,* November 1862, 527.

"Improvement in Hot-air Engines," *Scientific American,* April 24, 1869, 257.

"Is a Flying Machine a Mechanical Possibility?" *Scientific American,* March 13, 1869, 169.

"Panorama of Early Wings," *American Helicopter,* December 1945, 30.

"Petroleum as Fuel," *Scientific American,* December 2, 1893, 358.

Santos-Dumont, Alberto. "The Pleasures of Ballooning," *The Independent,* June 1, 1905, 1225.

"Steam Tried for Planes," *New York Times,* August 18, 1935, 7.

Taylor, Michael J. *History of the Helicopter.* London: Hamlyn Publishing, 1984.

"The Texas Beaumont Oil Well," *Scientific American,* Feb. 2, 1901, 74.

"The Great Balloon Voyage," *Ohio Farmer,* July 16, 1859, 228.

"The Philosophy of Balloons," *The Albion,* November 7, 1863, 533.

Thurston, R. H. "Steam and Its Rivals," *Forum,* May 1888, 341.

Chapter Four

"11 German Balloons His Bag in 4 Days," *New York Times,* September 18, 1921, 9.

"A Wingless Machine That Promises to Revolutionize Aerial Navigation," *Current Opinion,* March 1920, 407.

Aerofiles, "Powerplants—Reciprocating Engines," at http://www.aerofiles.com/motors.html (Accessed June 26, 2005).

Aerofiles, "Berliner—Berliner Joyce," at http://www.aerofiles.com/berlin.html (Accessed October 10, 2005).

"Airplane Ambulances," *Outlook,* May 14, 1919, 60.

Allen, Catherine. 2005. Personal communications. October–November.

"Asserts Mastery of Vertical Flight," *New York Times,* January 19, 1924, 3.

"Balloons the Feature of Armory Auto Show," *New York Times,* January 14, 1906, 10.

Barnard, Charles. "The Red Cross," *The Chautauquan,* December 1888, 143.

Berliner, Emile. "The Berliner Helicopter," *Aeronautics,* November 1908, 9.

"Berliner on His Aerobile," *New York Times,* September 9, 1908, 1.

Bracke, Albert (translated by Bernard Mettier). *Les Hélicoptères Cornu.* Paris: Librairie des Sciences Aéronautique, 1908.

Brown, Mike, 2006. Personal communication. 17 January.

De Bothezat, George. "The Meaning for Humanity of the Aerial Crossing of the Ocean," *Scientific Monthly,* November 1919, 433.

De Transehe, N. "Figure of Merit," *American Helicopter,* January 1947, 11.

De Villermont, Henri A. "French Rotary Wings," *American Helicopter,* May 1947, 10.

———. "Presenting…Louis Bréguet," *American Helicopter,* September 1947, 16.

Doherty, Trafford. 2006. Personal communication. 13 June.

"E. Berliner Dies; Famous Inventor," *New York Times,* August 4, 1929, 24.

"Einstein Disputes Lecturing Savant," *New York Times,* January 16, 1935, 19.

"France Buys Right to New Helicopter," *New York Times,* January 28, 1921, 8.

"Glenn Curtiss Dies, Pioneer in Aviation," *New York Times,* July 24, 1930, 1.

Gorn, Michael H. *The Universal Man: Theodore von Kármán's Life in Aeronautics.* Washington, D.C.: Smithsonian Institution Press, 1992.

Grosvenor, Edwin S. *Alexander Graham Bell: The Life and Times of the Man Who Invented the Telephone.* New York: Harry N. Abrams, 1997.

"Helicopter Ascends 7 Feet at Trials," *New York Times,* June 17, 1922, 2.

"Helicopter Maker, de Bothezat, Dead," *New York Times,* February 3, 1940, 11.

"Helicopter Opens New Flying Era," *New York Times,* July 15, 1923, 11.

History Office, Aeronautic Systems Center. *Splendid Vision, Unswerving Purpose: Developing Air Power for the United States Air Force During the First Century of Powered Flight.* Wright-Patterson AFB: Air Force History and Museums Program, 2002.

House, Kirk W. *Hell-Rider to King of the Air: Glenn Curtiss' Life of Innovation.* Warrendale, PA: SAE International, 2003.

Howland, Harold J. "The Sons of Daedalus," *Outlook,* September 26, 1908, 153.

Irwin, Will. "A Night Ride with the American Ambulance Corps at Verdun," *Current Opinion,* October 1916, 286.

Jane, Fred T. (ed.). *All the World's Aircraft, A Reprint of the 1913 Edition.* New York: Arco Publishing, 1969.

————. *All the World's Airships, A Reprint of the 1909 Edition.* New York: Arco Publishing, 1969.

Johnson, Thomas M., and Fletcher Pratt. *The Lost Battalion.* Lincoln: University of Nebraska Press, 2000.

Kruckman, Arnold. "Hammondsport an Aeroplane Laboratory," *Outing,* August 1910, 535.

"L'Hélicoptère Pescara," http://lehen.david.neuf.fr/helicopter.html (Accessed February 20, 2006).

Lame, Maurice. "French-Built Helicopters," *American Helicopter,* April 1947, 6.

Leishman, Gordon. "The Cornu Helicopter—First in Flight?" *Vertiflite,* Fall 2001, 54.

"Louis Bréguet, 75, Aircraft Pioneer," *New York Times,* May 5, 1955, 33.

"Louis Brennan Dead; British Inventor," *New York Times,* January 20, 1932, 19.

Lynde, Francis. "Soldiers of Rescue," *Outlook,* October 23, 1918, 294.

"Moors Kill Two Airmen," *New York Times,* April 9, 1914, 4.

Munk, Max M. *Techinical Note No. 221. Model Tests on the Economy and Effectiveness of Helicopter Propellers.* Washington, D.C.: National Advisory Committee for Aeronautics, 1925.

Paris Office. *Technical Memorandum No. 13. The Oehmichen–Peugeot Helicopter.* Washington, D.C.: National Advisory Committee for Aeronautics, 1931.

"Pioneer Helicopter Flies with Two Passengers," *Current Opinion,* April 1, 1923, 475.

"Predicts Helicopters Will Guard London," *New York Times,* October 4, 1925, 30.

Ramakers, L. "The Hélicoptère: Santos-Dumont's Latest Flying Machine," *Scientific American,* February 10, 1906, 129.

Rankine, W. J. M. "On the Mechanical Principles of the Action of Propellers," *Transactions of the Institute of Naval Architects,* 1865, 13.

"Rival Aviators All Over Europe," *New York Times,* March 28, 1909, 82.

Roseberry, C. H. *Glenn Curtiss: Pioneer of Flight.* Garden City, NJ: Doubleday & Co., 1972.

"Russian Invented Helicopter," *New York Times,* January 28, 1923, 5.

"Says He Has Craft to Fly Vertically," *New York Times,* March 8, 1920, 3.

"Says Helicopter Goes 312 Miles an Hour," *New York Times,* July 7, 1921, 1.

Shulman, Seth, *Unlocking the Sky: Glenn H. Curtiss and the Race to Invent the Airplane.* New York: HarperCollins, 2002.

Simanaitis, Dennis. 2006. Personal communication. 17 January.

"Sound Etched on Zinc: Electrician Berliner Has an Invention He Calls the Gramophone," *New York Times,* October 24, 1890, 6.

Studer, Clara. *Sky Storming Yankee: The Life of Glenn Curtiss.* New York: Arno Press, 1972.

"U.S. Navy Blimp Saves Canadians in Jungle," *New York Times,* March 25, 1944, 4.

"Vatican Orders 3 Helicopters," *New York Times,* December 19, 1930, 1.

Von Kármán, Theodore. *Technical Note. No. 47. Recent European Developments in Helicopters.* Washington, D.C.: National Advisory Committee for Aeronautics, 1921.

————. *The Wind and Beyond: Theodore von Kármán, Pioneer in Aviation and Pathfinder in Space.* Boston: Little, Brown and Co., 1968.

Warner, Edward P. *Technical Memorandum No. 107: The Prospects of the Helicopter.* Washington, D.C.: National Advisory Committee for Aeronautics, June 1922.

Wicks, Frank. "Trial by Flyer," *Mechanical Engineering,* 2003, 4.

"Will Fly Helicoptically: At Least Emile Berliner Intends to Try It in Washington Soon," *New York Times,* September 5, 1908, 2.

Williams, Hale P. "Beating the Bird in Its Own Realm," *Illustrated World,* November 1921, 909.

Williams, James. 2005. Personal communication. 11 December.

Chapter Five

"Aerial Study Made of Traffic Snarls," *New York Times,* June 24, 1935, 19.

"Air Transport Interested in Autogiro Landing," *New York Times,* January 3, 1932, 6.

"Army Buys 6 Autogiros," *New York Times,* March 3, 1937, 16.

"Asserts Boomerangs Were the First Autogiros," *New York Times,* January 10, 1931, 8.

"Autogiro Flies Once Again—In the Courts," *Business Week,* May 27, 1967, 78.

Brooks, Peter W. *Cierva Autogiros: The Development of Rotary-Wing Flight.* Washington D.C.: Smithsonian Institution Press, 1988.

"Buys a House for His Baby," *New York Times,* June 5, 1911, 1.

"Buys 'Autogyro' Rights; H. F. Pitcairn Hails New Ship an Return from Europe," *New York Times,* April 18, 1929, 24.

Charnov, Bruce H., 2005. Interview by author. 12 December.

————. *From Autogiro to Gyroplane: The Amazing Survival of an Aviation Technology.* Westport, CT: Praeger Publishers, 2003.

"Court Lets Stand Royalties Award for Autogiro," *Wall Street Journal,* January 24, 1978, 1.

Courtney, W. B. "A Latter-Day Pioneer," *Collier's,* September 12, 1931, 11.

De la Cierva, Juan. *Wings of Tomorrow: The Story of the Autogiro.* New York: Brewer, Warren & Putnam, 1991.

————. "Wings of Tomorrow." *Forum,* March 1931, 173.

Earhart, Amelia. "A Friendly Flight Across," *New York Times,* July 19, 1931, SM4.

————. *The Fun of It.* New York: Brewer, Warren and Putnam, 1932.

"Earhart to Receive Official Reprimand," *New York Times,* July 6, 1931, 3.

"Everything Went Black," *Time,* December 21, 1936, 20.

"Farley Dedicates 'Finest Postoffice,' " *New York Times,* May 26, 1935, 10.

"First Aircraft to Land Within World's Fair Grounds," *New York Times,* August 20, 1940, 22.

Fisher, Barbara E. Scott. "Notes of a Cosmopolitan," *North American Review,* January 1933, 97.

"Frees Mrs. Hopkins, but Censures Her," *New York Times,* April 6, 1915, 7.

" 'Giro Flies Mail a Year," *New York Times,* July 14, 1940, 119.

" 'Giro on Philadelphia–Camden Mail Run," *New York Times,* July 7, 1939, 1.

Gregory, H. F. *Anything a Horse Can Do: The Story of the Helicopter.* New York: Reynal & Hitchcock, 1944.

Hilton, R. "The Alleged Vulnerability of the Autogiro," *The Fighting Forces,* August 1934, 231.

Hirschberg, Michael, Thomas Müller, and Michael J. Pryce, "British V/STOL Rotorcraft in the Twentieth Century," *American Helicopter Society Forum 61,* June 2005.

" 'House Never Dark' Not Now Her Home," *New York Times,* December 24, 1914, 1.

"Hudson Pier Becomes Airport for Autogiro,"*New York Times,* December 24, 1931, 2.

"Juan de la Cierva," *New York Times,* December 12, 1936, 18.

"Juan de la Cierva, Spanish Loyalist," *New York Times,* January 13, 1938, 22.

Klemin, Alexander. "Learning to Use Our Wings," *Scientific American,* January 1926, 48.

Lopez-Diaz, C. Cuero-Rejado, and J. L. Lopez-Ruiz, in "Historical Rotorcraft Restoration: The C-30 Autogiro." *Proceedings of the Institution of Mechanical Engineers,* 1999, 71.

Macaulay, Neill. *The Sandino Affair.* Chicago: Quadrangle Books, 1985.

"Mansion a Christmas Gift," *New York Times,* December 7, 1911, 9.

"Married on a Yacht," *New York Times,* October 5, 1906, 7.

Martyn, T. J. C. "Autogiro Is Able to Land Upon Almost Any Backyard," *New York Times,* February 24, 1929, 141.

Miller, John. 2005. Interviews and personal communication with author. October–November.

"Miss Earhart Avoids Serious Autogiro Crash When Ship Fails to Rise in Abilene Take-Off," *New York Times,* June 13, 1931, 1.

Mitchell, William. "The Automobile of the Air," *Woman's Home Companion,* May 1932, 18.

"Mrs. Daniell Left $7 Million Estate," *New York Times,* March 14, 1928, 9.

Polmar, Norman. "Historic Aircraft: The Sea Services' First Rotary-Wing Aircraft," *Naval History,* September–October 1998, 53.

Polt, Richard. 2005. Personal communication. 1 November.

Ray, James G. "Straight Up," *Saturday Evening Post,* November 8, 1938, 14.

Sikorsky, Igor. *The Story of the Winged-S.* New York: Dodd, Mead and Co., 1967.

"Situation in Spain," *Wall Street Journal,* July 30, 1909, 3.

Smith, Frank Kingston. *A Legacy of Wings: The Harold F. Pitcairn Story.* New York: Jason Aronson, 1985.

"The Government and Autogiros," *Wall Street Journal,* April 27, 1938, 3.

"The Marine Autogiro in Nicaragua," *Marine Corps Gazette,* February 1953, 56.

"Trip in Fixed-Wing Plane Fatal to Giro Inventor," *Newsweek,* December 19, 1936, 28.

Trowbridge, John Townsend. *The Vagabonds and Other Poems.* Boston: Fields, 1869.

"Two Autogiro Bills Approved," *New York Times,* June 10, 1938, 11.

"Urge Development of the Dirigible." *New York Times,* May 18, 1931, 5.

"U.S. Seizes Recluse as Big Tax Evader," *New York Times,* February 6, 1960, 4.

White, Frank Marshall. "The Black Hand in Control in Italian New York," *Outlook,* August 16, 1913, 857.

" 'Windmill' Plane Flies Channel," *New York Times,* September 19, 1928, 1.

"Yancey Lands Autogiro in Ruins of Yucatan," *New York Times,* February 5, 1932, 3.

Chapter Six

"B. P. Labensky Dies," *New York Times,* October 26, 1950, 31.

Berry, M. "Flettner-282," *American Helicopter,* June 1947, 18.

Brady, Bob. 2005. Interview by author. 13 October.

Coates, Steve, and Jean-Christophe Carbonel. *Helicopters of the Third Reich.* Crowborough, UK: Classic, 2002.

Crider, John R. "Along the Far-Flung Airways—Helicopters Stir Study," *New York Times,* March 5, 1939, 11.

Delear, Frank J. *Igor Sikorsky: His Three Careers in Aviation.* New York: Dodd, Mead and Co., 1969.

Focke, Henrich. "The Focke Helicopters," *American Helicopter,* January 1947, 14.

"Fonck Plane Burns, 2 Die," *New York Times,* September 22, 1926, 1.

Francis, Devon. *The Story of the Helicopter.* New York: Coward-McCann, 1946.

Gandt, Robert L. *China Clipper: The Age of the Great Flying Boats.* Annapolis: Naval Institute Press, 1991.

Harris, Benjamin Hooper. 2005. Interview by author. 6 February.

"Helicopter Record Is Set by Sikorsky," *New York Times,* May 7, 1941, 19.

Hunt, William E. *Helicopter: Pioneering with Igor Sikorsky.* Shrewsbury, UK: Airlife Publishing, 1998.

Keogan, Joseph. *The Igor I. Sikorsky Aircraft Legacy.* Stratford, CT: Igor I. Sikorsky Historical Archives, 2003.

Kretvix, Bob. 2005. Interview by author. 13 October.

Lawrence, Thomas H. "The Sikorsky R-4 Helicopter," *Advanced Materials & Processes,* August 2003, 57.

Leishman, Gordon. "The Gyroplanes, Helicopters, and Convertiplanes of Raoul Hafner," *American Helicopter Society Forum 61,* June 5, 2005.

LePage, Wynn L. *Growing Up with Aviation.* Ardmore, PA: Dorrance, 1981.

Libertino, Dan. 2005. Interviews and personal communication with author. October–November.

Morris, Charles Lester. *Pioneering the Helicopter.* New York: McGraw-Hill, 1945.

Nachlin, Harry. 2006. Interviews by author. January–February.

"Nash-Kelvinator Teams Up with United Aircraft to Build Sikorsky Helicopters for U.S. Army," *Wall Street Journal,* July 7, 1943, 5.

Niland, James A. "Fifth Anniversary—First Cross Country Flight," *American Helicopter,* June 1947, 15.

O'Brien, Kevin. 2005. Interview by author. 6 February.

Reitsch, Hanna. *The Sky My Kingdom.* Novato, CA: Presidio Press, 1991.

"Russian Refugees' Plane, with 9 Aboard, Wrecked on Golf Links in Test Flight," *New York Times,* May 5, 1924, 17.

Shalett, Sidney. "New Air Weapons Pass Imagination," *New York Times,* December 31, 1942, 8.

Sikorsky, Igor I. "Technical Development of the VS-300 Helicopter During 1941," *Journal of the Aeronautical Sciences,* June 1942, 309.

————. "The Coming Air Age—As Told to Frederick C. Painton," *Atlantic Monthly,* September 1942, 33.

————. "Wings for Your Family!" *American Magazine,* March 1953, 41.

Sikorsky, Sergei. 2005. Interview by author. 28 December.

"Sikorsky Building Big Plane with Aid of Russian Exiles," *New York Times,* October 7, 1923, 5.

Smithsonian Air and Space Museum, "Platt-LePage XR-1," at http://www.nasm.si.edu/research/aero/aircraft/platt-le page xr-1.htm (Accessed September 2005).

"The Government and Autogiros," *Wall Street Journal,* April 27, 1938, 3.

Transcript, Russell Halligan Oral History Interview Summer 1976, by Horace Waggoner. Springfield, IL: Sangamon State University Oral History Office.

"Uncle Igor and the Chinese Top," *Time,* November 16, 1953, 25.

U.S. Centennial of Flight Commission, "Henrich Focke—Fa 61," http://www.centennialofflight.gov/essay/Rotary/Focke/HE5.htm (Accessed April 4, 2006).

"United Aircraft Buys Sikorsky," *New York Times,* July 19, 1929, 18.

Williamson, Samuel T. "The Whirling Rise of Mr. Helicopter," *New York Times,* September 13, 1959, SM99.

Wilson, Eugene E. "The Most Unforgettable Character I've Met: Igor Sikorsky," *Reader's Digest,* December 1956, 105.

Chapter Seven

Anderson, Ross. "Traveling Back to the Future at the 1962 World's Fair," *Seattle Times,* December 31, 1999, 1.

"Army Ships Repair Planes Downed at Sea," *New York Times,* May 1, 1945, 12.

"Arthur M. Young, 1905–1995," at http://www.arthuryoung.com (Accessed September 5, 2005).

Balkin, John. 2005. Interview by author. 5 February.

Beard, Barrett Thomas. *Wonderful Flying Machines: A History of Coast Guard Helicopters.* Annapolis, MD: Naval Institute Press, 1996.

Bierman, John, and Colin Smith. *Fire in the Night: Wingate of Burma, Ethiopia, and Zion.* New York: Random House, 1999.

Bradbury, Richard. "The RAF's Helicopters," *American Helicopter,* April 1946, 30.

Bridge, John. "Kiefer's Filling Station Pumps Super-Service at Rate of $100,000 Yearly," *Wall Street Journal,* September 10, 1948, 1.

Briscoe, C. H. "Helicopters in Combat: World War II," *Special Warfare,* Summer 2001, 32.

Carle, Louis. "I Flew Them in Combat," *American Helicopter,* January 1947, 10.

Deigan, Edgar. "The Flying Bananas and How They Grew," *Flying,* July 1949, 24.

"Detroit's Air Taxi," *Business Week,* December 11, 1943, 30.

Dorr, Robert F. *Chopper: Firsthand Accounts of Helicopter Warfare World War II to Iraq.* New York: Berkley Books, 2005.

Fergusson, Bernard. *Beyond the Chindwin: Being an Account of the Adventures of Number Five Column of the Wingate Expedition into Burma.* London: Collins, 1943.

Floherty, John J., and Mike McGrady. *Whirling Wings.* Philadelphia: J. B. Lippincott, 1961.

Harman, Carter. "Mission in Burma," *American Helicopter,* March 1946, 17.

Harris, Carl. 2002. Interview by author. 10 February.

"Heads Plane Company at 22," *New York Times,* November 1, 1947, 2.

"He Beats the Traffic by Helicopter," *Business Week,* April 21, 1956, 114.

"Helicopter from Here Saves Fliers Marooned in Labrador Wilderness," *New York Times,* May 4, 1945, 21.

"Helicopter Lands with Saucy Hello," *New York Times,* April 13, 1949, 34.

"Helicopter Rescues Boy," *New York Times,* April 4, 1944, 23.

"Helicopter Rushes Plasma in History-Making Flight," *Wall Street Journal,* March 8, 1944, 7.

"Helicopters Which Can Compete with Surface Transportation Seen by July," *Wall Street Journal,* October 19, 1945, 2.

"Hillercopter," *Time,* September 11, 1944, 55.

Holder, John. "Eggbeater: Kellett's Radical Rotor Head Design," *American Helicopter,* January 1946, 36.

Konke, Curt. 2005. Interview by author. 5 February.

Larsen, Agnew E., and Joseph S. Pecker. "What Is the Helicopter's True Commercial Future?" *Aviation,* December 1943, 116.

"Lawrence Bell, Air Leader, Dead," *New York Times,* October 21, 1956, 87.

Leary, William. "The Helicopter Goes to Sea," *Flying,* September 1949, 26.

Leavitt, Lou. "Let's Be Calm About the Helicopter," *Aviation,* November 1943, 114.

Lert, Peter. "Whatever Happened to 'A Helicopter in Every Garage'?" *Air Progress,* November 1978, 40.

Macrae, Ray. 2005. Interview by author. 5 February.

Mashman, Joe. *To Fly Like a Bird, As Told to R. Randall Padfield.* Potomac, MD: Phillips Publishing, 1992.

McGinley, Phyllis. "All God's Chillun Got Helicopters," *The New Yorker,* August 21, 1943, 24.

Nagy, Barbara A. "Inventor, Musician, Businessman, Samaritan," *Hartford Courant,* November 17, 1997, D10.

"New Flying Machine," *Time,* March 8, 1943, 51.

Norton, Donald J. *Larry: A Biography of Lawrence D. Bell.* Chicago: Nelson-Hall, 1981.

"Personal Aircraft," *Business Week,* September 26, 1942, 13.

"Piasecki: Getting Set for Mass Transportation," *Business Week,* September 26, 1953, 144.

Polmar, Norman. "The Amazing Hup-Mobile," *Naval History,* May–June 1999, 61.

"Presenting . . . Stanley Hiller Jr.," *American Helicopter,* July 1947, 15.

"PV-2 Makes Public Flight," *Aviation,* November 1943, 229.

"Rotor Aircraft, Wingless, Is Tested," *New York Times,* August 20, 1930, 3.

Salpukas, Agis. "Arthur M. Young Dies at 89," *New York Times,* June 3, 1995, 11.

Sanduski, John J. "Pacific Venture," *American Helicopter,* December 1946, 13.

Senderoff, Izzy. 2005. Interview by author. 10 October.

"Sikorsky Hovers in Air with Greatest Ease and Sells Army on Future of His Helicopter," *Newsweek,* March 8, 1943, 58.

Spenser, Jay P. *Vertical Challenge: The Hiller Aircraft Story.* Seattle: University of Washington Press, 1992.

———. *Whirlybirds: A History of the U.S. Helicopter Pioneers.* Seattle: University of Washington Press, 1999.

Straubel, John F. *One Way Up.* Palo Alto, CA: Hiller Aircraft Co., Division of Fairchild Hiller, 1964.

Taylor, Frank J. "Look at the Tricks He Does in the Air!" *Saturday Evening Post,* June 28, 1952, 25.

"The K-190: New Helicopter Is Maneuverable, Safe and a Perfect Cinch to Fly," *Life,* November 15, 1948, 63.

Thompson, Julian. *The Imperial War Museum Book of the War in Burma 1942–1945.* London: Pan Books, 2002.

Tipton, Richard S. *They Filled the Skies.* Fort Worth, TX: Bell Helicopter Textron, 1989.

Vandercrift, John L. *A History of the Air Rescue Service.* Winter Park, FL: Rollins Press, 1959.

Veazey, Robert. 2004. Interview by author. 13 May.

Wales, George. "Prepared for the Future," *American Helicopter,* April 1947, 19.

Wambold, Donald. "Frank Piasecki Left Behind an Enduring Legacy in the Innovative PV-2," *Aviation History,* July 2005, 64.

"West Coast Lad, 19, Improves Helicopter, Flies Own Ship Without Tail Propeller," *New York Times,* August 31, 1944, 19.

"Will Try Atlantic with Sailless Ship," *New York Times,* November 7, 1924, 1.

Williams, Gurney. "Park on a Cloud," *Collier's,* May 15, 1943, 14.

Chapter Eight

Adams, Claude D. "An Idea Was Born," *American Helicopter,* August 1947, 14.

"At $2,000,000 Cost Queens Gets Roads," *New York Times,* September 12, 1912, 8.

Balchen, Bernt. *The Next Fifty Years of Flight.* New York: Harper, 1954.

Baum, Dale, and James L. Hailey. "Lyndon Johnson's Victory in the 1948 Texas Senate Race," *Political Science Quarterly,* Fall 1994, 595.

Bunkley, Allison W. "The Test," *American Helicopter,* October 1946, 24.

Caro, Robert A. *Means of Ascent.* New York: Alfred A. Knopf, 1990.

Carroll, George. "A Helicopter on Every Roof?" *American Mercury,* 81.

Carroll, Ruth, and Latrobe Carroll. *The Flying House.* New York: Macmillan, 1946.

Cooke, Richard P. "The Helicopter Is Practical," *Wall Street Journal,* May 10, 1943, 1.

Dallek, Robert. *Lone Star Rising: Lyndon Johnson and His Times, 1908–1973.* New York: Oxford Institute Press, 1991.

Edgar, Norman. "It's in the Hat," *American Helicopter,* December 1946, 29.

Enstrom, Edith. 2007. Interview by author. 12 January.

"Gay Plot of Times Fashion Show Adds Drama to Newest Style," *New York Times,* November 1, 1946, 25.

Gemmill, Henry. "A Young Yank Soars from Los Angeles to Borneo and Bengasi," *Wall Street Journal,* July 12, 1956, 1.

"Gigantic Airplane Carried 16 Persons Successfully," *New York Times,* April 5, 1914, SM3.

Goddard, Stephen. *Getting There: The Epic Struggle between Road and Rail in the American Century.* New York: Basic Books, 1994.

Hart, Dick. 2006. Interview by author. 16 January.

"Helicopter Air-Bus Service Planned by Greyhound Corp.," *Wall Street Journal,* June 16, 1943, 4.

"Helicopter Air Transport Company Pioneers a New Field in Aviation," *Flying,* August 1947, 23.

"Helicopter Hubbub: Many Young Inventors Try for Mass Market," *Wall Street Journal,* September 17, 1947, 1.

"Helicopter Panorama 1946," *American Helicopter,* December 1946, 16.

"Helicopter Program Lags Behind Needs," *Aviation Week,* March 12, 1956, 258.

"Helicopters Inc. to Quit: Officials Report No Market for Product," *New York Times,* August 25, 1949, 35.

"Hermit Hops in Helicopter," *American Helicopter,* April 1947, 37.

"Higgins Reveals Plans for Cheap Helicopter," *Wall Street Journal,* January 28, 1943, 6.

"Hiller HOE-1 / YH-32 'Hornet,' " at http://avia.russian.ee/vertigo/hiller hoe-1-r.html (Accessed January 10, 2006).

Holt, W. J. "He Likes to Fly Straight Up," *Saturday Evening Post,* August 11, 1951, 32.

Johnson, Sam Houston. *My Brother Lyndon.* New York: Cowles, 1969.

"Johnson Lashes Civil Rights and GOP in Talk Here," *Terrell* (TX) *Tribune,* June 15, 1948, 1.

"Johnson's 'Flying Windmill' Unique Political Craft," *Terrell* (TX) *Tribune,* June 15, 1948, 1.

Klemin, Alexander. "The Problem of the Helicopter," *Scientific Monthly,* August 1948, 127.

——. *The Helicopter Adventure.* New York: Coward-McCann, 1947.

Lenhardt, Jack. 2006. Interview by author and personal communication. 21 February.

Macauley, C. B. F. *The Helicopters Are Coming.* New York: McGraw-Hill, 1944.

Miller, Merle. *Lyndon: An Oral Biography.* New York: Putnam, 1980.

Patton, Phil. *Open Road.* New York: Simon & Schuster, 1986.

Pierce, Bert. "President Decries 'Nuts' Driving Cars," *New York Times,* May 9, 1946, 1.

"Plans Helicopter Bus Lines," *New York Times,* June 15, 1943, 13.

Reed, William S. "Los Angeles Helicopter Utilization Grows," *Aviation Week & Space Technology,* March 5, 1962, 57.

"Rotary Wing Aircraft," *Flying,* November 1961, 27.

Salisbury, Harrison. "Study Finds Cars Choking Cities as 'Urban Sprawl' Takes Over," *New York Times,* 1959, 1.

Smith, H. Alexander. *Diary, 1946* (Unpublished document at Seeley G. Mudd Manuscript Library, Princeton University).

Sullivan, Mark. *Our Times.* New York: Scribner, 1926.

"The Helicopter: A War Baby with a Big Future," *Newsweek,* September 20, 1954, 80.

"Trade to Test Helicopter," *New York Times,* August 18, 1944, 11.

Transcript, Horace Busby Oral History Interview II, 3/4/82, by Michael L. Gillette. Austin, TX. LBJ Library. At http://www.lbjlib.utexas.edu/johnson/archives.hom/oralhistory.hom/BusbyH/Busby2.pdf (Accessed March 19, 2006).

Transcript, James E. Chudars Oral History Interview I, 10/2/1981, by Michael L. Gillette. Austin, TX. LBJ Library. At http://www.lbjlib.utexas.edu/johnson/archives.hom/oralhistory.hom/Chudars-j/Chudars.PDF (Accessed March 21, 2006).

Transcript, Joe Mashman Oral History Interview I, by Joe B. Frantz, Internet Copy, LBJ Library. Austin, TX. LBJ Library, at http://www.lbjlib.utexas:edu/johnson/archives.hom/oralhistory.hom/Mashman-J/Mashman1.PDF (Accessed March 20, 2006).

Van Lopik, Carter. "Neighbors Bankroll U.P. Helicopter Firm," *Detroit News,* January 26, 1964, 1.

White, Peter T. "The Incredible Helicopter," *National Geographic,* April 1959, 533.

"Wide Use of Helicopters Predicted After the War," *Wall Street Journal,* April 10, 1943, 1.

Chapter Nine

"2 Colonels Killed in Collision of Helicopters Near Saigon," *New York Times,* September 19, 1969, 14.

"Aeromedical Evacuation," *Air Power History,* Summer 2000, 38.

"Air Force Orders Giant Helicopter," *New York Times,* October 3, 1948, 14.

André, Valérie. *Remarks by Medicine General Inspector Valerie Andre at Whirly-Girls 30th Anniversary Dinner.* (Unpublished manuscript, Texas Women's University Collection.)

Baldwin, Hanson W. "Wintershield II," *New York Times,* February 10, 1960, 3.

———. "War Game Aided by Huge Airlift," *New York Times,* April 30, 1954, 10.

Barrett, George. "Helicopter Unit Saved 200 in Korea," *New York Times,* February 26, 1951, 3.

Breuer, William. *Shadow Warriors: The Covert War in Korea*. New York: John Wiley & Sons, 1996.

Champlin, G. F. "New Ramjet 'Hiller Hornet,'" *American Helicopter*, February 1951, 8.

Cipalla, Rita. "Sky's No Limit," *Chicago Tribune*, March 29, 1987, 8.

Cole, Richard B. "Economically Glum New England Boasts Hartford Bright Spot," *Wall Street Journal*, March 1, 1952, 1.

Constant, Edward W. *Origins of the Turbojet Revolution*. Baltimore: Johns Hopkins University Press, 1980.

"Copter Lines Said to Face Failures," *New York Times*, December 12, 1964, 62.

Davis, John L. " 'Eggbeaters' Make Combat Debut in Korea," *Veterans of Foreign Wars Magazine*, January 2002, 40.

Delgado, James P. "Bombshell at Bikini," *Naval History*, July–August 1996, 33.

Dwiggins, Don. "Pinwheel Man," *Flying*, February 1952, 34.

"Enstrom Copter Story," *Herald-Leader* (Menominee, MI), January 5, 1968, 2.

Farrell, Robert. "French Meet Guerrillas with Helicopters," *Aviation Week*, September 17, 1956, 28.

———. "Algerian Terrain Challenges Helicopters," *Aviation Week*, September 24, 1956, 88.

Feron, James. "Big, Fast Israeli Copters Carry a Heavy Burden Under New Military Concept," *New York Times*, January 29, 1969, 14.

Francis, C. B. "The Seibel Helicopter," *American Helicopter*, August 1947, 17.

"French Destroy Rebel Fortress," *New York Times*, September 11, 1956, 14.

"Helicopter Parade Whirls Over Capital," *New York Times*, April 30, 1951, 29.

"Helicopter Production Must Be Increased, Army Report States," *Wall Street Journal*, March 11, 1954, 1.

Hershey, Burnet. *The Air Future: A Primer of Aeropolitics*. New York: Duell, Sloan and Pearce, 1943.

"Holiday Accidents Kill 793, a Record," *New York Times*, July 6, 1950, 29.

Horne, Alistair. *A Savage War of Peace*. New York: Viking Adult, 1978.

"How Hiller Tests Tiny Ramjet," *Aviation Week*, December 20, 1954, 4.

"Huge Helicopter Takes to the Air," *New York Times*, October 24, 1952, 48.

"Jet Helicopter Wrecked—Hughes' Giant XH-17 Rips Loose from Moorings," *New York Times*, June 23, 1950, 36.

Johnston, Richard J. H. "Marines 'Attack' (At Own Expense)," *New York Times,* March 21, 1950, 12.

Kocks, Kathleen. "Helicopters Hunters, Not Victims," *Journal of Electronic Defense,* February 2000, 33.

Kreisher, Otto. "Rocks and Ridgetops," *Sea Power,* June 2001, 53.

Lessing, Lawrence P. "Helicopters," *Scientific American,* January 1955, 37.

Marion, Forrest L. "The Grand Experiment: Detachment F's Helicopter Combat Operations in Korea, 1950–1953," *Air Power History,* Summer 1993, 38.

McAllister, G. J. "Army Reviews Copter Lessons," *Aviation Week,* March 22, 1954, 25.

McGregor, Greg. "Heartbreak Ridge Completely Won by Allied Assault," *New York Times,* October 12, 1951, 1.

"Mlle. le Docteur Annoys Vietminh," *New York Times,* August 10, 1952, 4.

Montross, Lynn. *Cavalry of the Sky: The Story of U.S. Marine Combat Helicopters.* New York: Harper & Bros., 1954.

"Present Helicopters Inadequate: Army," *Aviation Week,* May 10, 1954, 17.

Schlaifer, Robert. *Development of Aircraft Engines.* Boston: Harvard University/Maxwell Reprint Co., 1950.

Shipp, Warren, and Howard Levy. "Rotary Wing Review," *Flying,* May 1952, 29.

Shrader, Charles R. *The First Helicopter War: Logistics and Mobility in Algeria, 1954–1962.* Westport, CT: Praeger Publishers, 1999.

Sulzberger, C. L. "Trouble Ahead with Paris—Helicopter Headaches," *New York Times,* June 18, 1955, 16.

"Taming the Whirly-Bird," *Flying,* November 1951, 74.

"Test Equipment Failure Believed Cause of Piasecki YH-16A Crash," *Aviation Week,* January 16, 1956, 34.

"Up from the Basement," *Newsweek,* January 20, 1964, 70.

"Up with the Helicopter," *Fortune,* May 1951, 91.

Van Lopik, Carter. "Neighbors Bankroll UP Helicopter Firm," *Detroit Free Press,* January 26, 1964, 1.

"Vertical Envelopment," *Flying,* November 1951, 63.

Weeghman, Richard B. "Pilot Report: The Enstrom," *Flying,* September 1968, 68.

Witze, Claude O. "Helicopter Builders Grapple with Cost," *Aviation Week,* July 5, 1954, 13.

Chapter Ten

Apple, R. W. "Copter Division Reaches Vietnam," *New York Times,* September 13, 1965, 1.

Arnett, Peter, "After Two Years in Mekong Delta, U.S. Goal Is Elusive," *New York Times,* April 15, 1969, 12.

Baker, Russell. "Long Slow Fight in Vietnam Seen," *New York Times,* May 1, 1962, 15.

Barber, Noel. *The War of the Running Dogs: The Malayan Emergency: 1948–1960.* New York: Weybright and Talley, 1971.

Bradin, James W. *From Hot Air to Hellfire: The History of Army Attack Aviation.* Novato, CA: Presidio, 1994.

Brown, Russell K. "Fallen Stars," *Military Affairs,* February 1981, 9.

Burchett, Wilfred G. *Vietnam: The Inside Story of the Guerilla War.* New York: International Publishers, 1965.

Chanoff, David, and Doan Van Toai. *Portrait of the Enemy.* New York: Random House, 1986.

Chapelle, Dickey. "Helicopter War in South Viet Nam," *National Geographic,* November, 1962, 722.

Chinnery, Philip D. *Vietnam: The Helicopter War.* Annapolis: Naval Institute Press, 1991.

Cleveland, Les. "Songs of the Vietnam War: An Occupational Folk Tradition," *New Directions in Folklore* (2003). Internet Journal, at http://www.temple.edu/isllc/newfolk/military/songs.html (Accessed February 1, 2007).

Conboy, Ken. "Early Covert Action on the Ho Chi Minh Trail," *Vietnam,* August 2000, 30.

Cooke, Richard P. "Bigger, Faster Craft For Civilian Use Grow Out of Military Needs," *Wall Street Journal,* November 8, 1965, 1.

Doleman, Edgar C. *Tools of War.* Boston: Boston Publishing, 1985.

Dooley, George. "17 Years in Vietnam," *Vietnam,* February 2007, 37.

Dougherty, Kevin J. "The Evolution of Air Assault," *Joint Forces Quarterly,* Summer 1999, 51.

Easterbrook, Gregg. "All Aboard Air Oblivion," *Washington Monthly,* September 1981, 14.

Emerson, Henry E. 2006 and 2007. Interviews by author. February and January.

———. *Can We Out-Guerilla the Communist Guerillas?* Carlisle, PA: U.S. Army War College, 1965.

320   BIBLIOGRAPHY

"Enemy Fire Kills U.S. General, the Fifth to Die in Vietnam War," *New York Times,* April 2, 1970, 3.

Ennis, John. "Helicopters: Unsafe at Any Height," *Popular Mechanics,* September 1971, 63.

Ewell, Julian. 2006. Interview by author. 5 March.

Ewell, Julian J., and Ira A. Hunt. *Sharpening the Combat Edge: The Use of Analysis to Reinforce Military Judgment.* Washington, D.C.: Department of the Army, Superintendent of Documents, 1974.

Fall, Bernard. *Street Without Joy.* Harrisburg, PA: Stackpole, 1961.

"G.I.'s Use Hatchets in a Jungle Fight Against Vietcong," *New York Times,* December 13, 1965, 5.

Giap, Vo Nguyen. *People's War, People's Army* (translation). Washington, D.C.: Department of Defense, 1962.

"Good-Luck Hank," *Newsweek,* September 9, 1968, 56.

Grau, Lester W. "The RPG-7: On the Battlefields of Today and Tomorrow," *Infantry,* May–August 1998, 6.

Grimes, Paul. 2006. Interview by author. 21 February.

Gurney, Gene. *Vietnam, the War in the Air.* New York: Crown, 1985.

Halberstadt, Hans. *Army Aviation.* Novato, CA: Presidio Press, 1990.

Hay, John H. *Vietnam Studies: Tactical and Material Innovations.* Washington: Department of the Army, 1989. At www.army.mil/cmh/books/vietnam/tactical/chapter2.htm (Accessed June 30, 2006).

Heuer, Marty. 2005. Personal communication. December.

Hickey, Gerald C. 2007. Interview by author. 19 February.

———. *Free in the Forest: Ethnohistory of the Vietnamese Central Highlands, 1954–1976.* New Haven, CT: Yale University Press, 1982.

———. *Village in Vietnam.* New Haven, CT: Yale University Press, 1964.

———. *Window on a War: An Anthropologist in the Vietnam Conflict.* Lubbock, TX: Texas Tech University Press, 2002.

"Just Say It Was the Comancheros," *Newsweek,* March 15, 1971, 39.

Karnow, Stanley. "Giap Remembers," *New York Times,* June 24, 1990, SM22.

Lengyel, Greg. 2007. Interview by author and personal communication. 9 February.

Lindsay, James. 2007. Interview by author. 31 January.

Lloyd, Barry. 2006. Interviews and personal communication with author. January and June.

Lundh, Lennart. *Sikorsky H-34: An Illustrated History.* Chicago: Schiffer Publishing, 1998.

Mason, Robert. *Chickenhawk.* New York: Viking Press, 1983.

"Mekong Delta Still Paralyzed 5 Weeks After Foe's Offensive," *New York Times,* March 8, 1968, 4.

Mertel, Kenneth D. *Year of the Horse: Vietnam–1st Air Cavalry in the Highlands, 1965–1967.* Atglen, PA: Schiffer Military/Aviation History, 1996.

Miers, Richard. *Shoot to Kill.* London: Faber and Faber, 1959.

Mills, Hugh L., and Robert A. Anderson. *Low-Level Hell: A Scout Pilot in the Big Red One.* Novato, CA: Presidio Press, 1992.

Mohr, Charles. "G.I.'s Fighting in Delta Use Stealth and Surprise," *New York Times,* May 22, 1968, 2.

———. "Radar Enables G.I.'s to Keep Close Eye on Enemy," *New York Times,* May 24, 1968, 5.

———. "Helicopters Save Lives in Vietnam," *New York Times,* February 20, 1966, 5.

Nevard, Jacques. "U.S. 'Copter Units Arrive in Saigon," *New York Times,* December 12, 1961, 21.

Orr, Kelly. *From a Dark Sky.* Novato, CA: Presidio Press, 1996.

Page, Tim. *Another Vietnam: Pictures of the War from the Other Side.* Washington, D.C.: National Geographic, 2002.

Penchenier, Georges. "Close-Up of the Vietcong in Their Jungle," *New York Times,* September 13, 1964, SM27.

Plaster, John L. *SOG: The Secret Wars of America's Commandos in Vietnam.* New York: Simon & Schuster, 1997.

Pribbenow, Merle (translator). *Victory in Vietnam: The Official History.* Lawrence, KS: University Press of Kansas, 2002.

Raymond, Jack. "In Zone D, Terrain is Snipers' Ally," *New York Times,* June 30, 1965, 1.

———. "Army to Increase Helicopter Force," *New York Times,* January 19, 1966, 5.

———. "It's a Dirty War for Correspondents, Too," *New York Times,* February 13, 1966, 219.

"Red Force Overruns Hamlet in Vietnam," *New York Times,* August 21, 1963, 1.

Reporting Vietnam. New York: Library of America, 1998.

Richards, Brien. 2006. Interview by author and personal communications. February.

Roberts, Gene. "Marines Advance in Hue," *New York Times,* February 7, 1968, 1.

———. 2006. Interview by author. 5 March.

Sheehan, Neil. *A Bright Shining Lie: John Paul Vann and America in Vietnam.* New York: Vintage Books, 1989.

Short, Anthony. *The Communist Insurrection in Malaya, 1948–1960.* London: Frederick Muller, 1975.

Siler, Charles. 2006. Interview by author. 11 March.

Simpson, Jay Gordon. "Not by Bombs Alone: Lessons from Malaya," *Joint Forces Quarterly,* Summer 1999, 91.

Sloniker, Mike. 2006. Personal communication. 10 January.

Smith, Hedrick. "Vietcong Terrorism Sweeping the Mekong Delta as Saigon's Control Wanes," *New York Times,* January 12, 1964, 14.

Smith, Tom. *Easy Target: The Long Strange Trip of a Scout Pilot in Vietnam.* Novato, CA: Presidio Press, 1996.

Steinbrunn, Robert. 2005. Interview by author. 14 July.

Story, Edward. 2005. Interviews and personal communication with author. 6 February.

Taylor, Thomas H. 2007. Interview by author. 29 January.

"The Bloody Checkerboard," *Newsweek,* May 23, 1966, 64.

Treaster, Joe. 2006. Interview and personal communication with author. March–June.

Trumbull, Robert. "How Communists Operate in Southeast Asia," *New York Times,* April 14, 1963, 161.

"U.S. 'Copters Rout Reds in Vietnam," *New York Times,* February 2, 1962, 1.

"Use of Medical Helicopters Raises Survival Rate of War Wounded," *New York Times,* May 21, 1967, 22.

Wells, R. (ed.). *The Invisible Enemy: Booby-traps in Vietnam.* Miami, FL: J. Flores Publications, 1992.

Chapter Eleven

"Air America: Played a Crucial Part of the Emergency Helicopter Evacuation of Saigon," http://www.historynet.com/wars conflicts/vietnam war/3035911.html (Accessed March 22, 2006).

Blumenthal, Ralph. "U.S. Copter Pilots Taking Some of Worst Fire of War," *New York Times,* February 12, 1971, 3.

Butler, David. *The Fall of Saigon.* New York: Simon & Schuster, 1985.

Butterfield, Fox, and Kari Haskell. "Getting It Wrong in a Photo," *New York Times,* April 23, 2000, 4.

Church, George J. "Saigon: The Final 10 Days," *Time,* April 24, 1995, 24.

Denman, Della. "They're Weaving a Story of War," *New York Times,* April 25, 1973, 32.

Dillon, Barry. "The Man Who Saved His Life," *Citizen Airman,* September 1988, 8.

Dunham, Mike. "Heroes in Our Midst: Alaska Vets Recall Epic Vietnam Battle Overlooked," *Anchorage Daily News,* September 17, 2000, H1.

Fisher, Gary. "Goodnight Saigon," *Leatherneck,* May 2005, 59.

Frisbee, John L. "A Tale of Two Crosses," *Air Force Magazine,* February 1992, 21.

Garland, Ed. 2006. Interview by author. 5 July.

Harnage, O. B. *A Thousand Faces.* Victoria, BC: Trafford Publishing, 2002.

Harrison, Benjamin L. *Hell on a Hill Top: America's Last Major Battle in Vietnam.* Lincoln, NE: iUniverse, 2004.

———. 2006. Interviews and personal communication with author. December.

Henderson, Charles. *Goodnight Saigon: The True Story of the U.S. Marines' Last Days in Vietnam.* New York: Berkley, 2005.

Knight, Wayne. 2006. Personal communication. 12 June.

LaPointe, Robert. *PJs in Vietnam: The Story of Air Rescue in Vietnam.* Anchorage, AK: Northern PI Press, 2002.

Middleton, Drew. "Army to Test New Triple-Threat Division Regarded as a Breakthrough in Land Warfare," *New York Times,* April 12, 1971, 27.

Morris, George. "Firebase Ripcord Veterans Recall Fierce Fighting Around Vietnam Mountaintop," *Advocate,* October 8, 2000, 14.

Nolan, Keith William. *Ripcord: Screaming Eagles Under Siege.* Novato, CA: Presidio Press, 2000.

"Saigon Copter Lands on Another in Stampede to U.S. Ship's Deck," *New York Times,* April 30, 1975, 85.

Sterba, James P. "13 Americans Die in Vietnam Clash," *New York Times,* July 23, 1970, 1.

"That Others May Live," *Time,* July 22, 1966, 27.

"The Invasion Ends," *Time,* April 5, 1971, 24.

"U.S. Copters Evacuate Periled Vietnam Village," *New York Times,* June 10, 1965, 2.

"U.S. Planes Blast an Abandoned Base Area Near Laos," *New York Times,* July 25, 1970, 3.

Walker, Fred. "The Fall of Saigon, April 1975," at http://www.air-america.org/Articles/Fall of Saigon.shtml (Accessed August 28, 2005).

Windrow, Martin. *The Last Valley: Dien Bien Phu and the French Defeat in Vietnam.* New York: Da Capo Press, 2004.

Woods, Chris. "Operation Frequent Wind," at http://www.fallofsaigon.org/woods.htm (Accessed August 28, 2005).

Chapter Twelve

"A Chopper Turns Deadly," *Newsweek,* May 30, 1977, 27.

Anderson, David. 2005. Interview by author. 5 February.

Bearak, Barry. "War Zone on Top of the World," *National Post,* May 24, 1999, A13.

Bearden, Milton. "Afghanistan, Graveyard of Empires," *Foreign Affairs,* November–December 2001, 17.

Bedell, Douglas. "Mini-Cars, Compact Buses Being Studied as Solutions to Traffic Snarls in Cities," *Wall Street Journal,* June 7, 1967, 7.

Berry, John M. "Seeking Oil in the West," *Washington Post,* October 5, 1980, G1.

Borovik, Artyom. *The Hidden War: A Russian Journalist's Account of the Soviet War in Afghanistan.* New York: Atlantic Monthly Press, 1990.

"Briton Flies Helicopter to Work," *New York Times,* June 28, 1947, 4.

Buder, Leonard. "A Way to Save People on Roof: Use Helicopters," *New York Times,* January 27, 1979, 23.

Bulloch, John, and Harvey Morris. *The Gulf War: Its Origins, History and Consequences.* London: Methuen, 1989.

Clines, Boyd. 2002. Interview by author. 10 February.

"Commercial Helicopters: They Need Subsidies to Fly," *Time,* May 16, 1955, 96.

"Construction of City's First Aerial 'Heliport' Begun Atop Port Authority's Headquarters," *New York Times,* November 28, 1950, 39.

"Convenient Roof Ports a Must for Economical Helicopter Flights," *Aviation Week,* August 22, 1955, 21.

Cook, Robert H. "Flying Crane Considered for Commuters," *Aviation Week,* February 15, 1960, 43.

Cooper, Tom. 2006. Personal communication. 16 February.

Cooper, Tom, and Farzad Bishop. *Iran-Iraq War in the Air.* Atglen, PA: Shiffer Military History, 2000.

Cordesman, Anthony H. *The Lessons of Modern War,* vol. 3. Boulder, CO: Westview Press, 1990.

Cordesman, Anthony, and Abraham R. Wagner. *The Lessons of Modern War: The Iran-Iraq War,* vol. 2. Boulder: Westview Press, 1990.

Crile, George. *Charlie Wilson's War.* New York: Atlantic Monthly Press, 2003.

Daley, Glenn. 2005. Interviews by author. August.

"Daring Rescue During Mill Fire," *Firehouse,* April 2000, 44.

Davis, Bill. 2002. Interview by author. 2 April.

Delear, Frank J. "Executive Helicopter Pilot," *Flying,* January 1961, 56.

"Dramatic Rescue in Hotel Disaster," *New York Times,* January 4, 1987, 22.

Edison, Thomas. "The Scientific City of the Future," *Forum,* December 1926, 823.

Edwards and Kelcey Engineering, Inc. *Heliport and Helicopter Master Plan for New York City, Final Report.* New York: City of New York, 1999.

Engel, Patrick. 2005. Personal communication. 25 June.

Everett-Heath, John. *Helicopters in Combat: The First Fifty Years.* New York: Sterling, 1992.

———. *Soviet Helicopters: Design, Development and Tactics.* London: Jane's Information Group, 1988.

Farrell, Robert. "Helicopter Taxis Buzz into Competition with the Earthbound Variety," *Wall Street Jounral,* July 12, 1951, 1.

Feerst, Bob. 2000 and 2005. Interview by author.

Finney, John W. "Iran Will Buy $2 Billion in U.S. Arms over the Next Several Years," *New York Times,* February 22, 1973, 2.

Fiszer, Michael. "The Mighty Mi-24," *Journal of Electronic Defense,* May 2005, 40.

"Flying Fire Apparatus Is Predicted by Kenlon," *New York Times,* June 14, 1929, 19.

"Flying Fire Engine Successful in First Public Test," *McDonnell Douglas Spirit,* September 1978, 2.

Friedlander, Paul J. C. "Midtown Launch Pad Opens Tuesday," *New York Times,* December 19, 1965, 1.

Gabbella, William. "Copter Commuting," *Flying,* May 1959, 33.

Gamauf, Mike. "Moving Mountains of Concreate of Muhammad: How Helicopters Were Key to the Building of Modern Jeddah," *Business & Commercial Aviation,* July 1, 2005, 66.

Gillies, Peter. 2006. Interviews and personal communication with author. January–February.

Girardet, Edward. "Afghan Guerillas Turning Soviet Attack to Their Advantage," *Christian Science Monitor,* May 31, 1984, 1.

———. "Afghan Guerilla Leader Holds His Own Against Soviet Offensive," *Christian Science Monitor,* October 2, 1984, 1.

———. *Afghanistan: The Soviet War.* New York: St. Martin's Press, 1985.

Graham, Frederick. "Handyman of the Skies," *New York Times,* December 3, 1950, SM14.

Grau, Lester W., and James H. Adams. "Air Defense with an Attitude: Helicopter v. Helicopter Combat," *Military Review,* January–February 2003, 22.

Grau, Lester (ed.). *The Bear Went Over the Mountain: Soviet Combat Tactics in Afghanistan.* Washington, D.C.: National Defense University, 1996.

Grau, Lester. 2006. Interviews and personal communication with author. August.

Griffiths, David R. "Iran Begins to Use Cobras, Mavericks," *Aviation Week & Space Technology,* October 13, 1980, 24.

Guilmartin, John F. *A Very Short War: The Mayaguez and the Battle of Koh Tang.* College Station, TX: Texas A&M Press, 1995.

Gunston, John. "Stingers Used by Afghan Rebels Stymie Soviet Air Force Tactics," *Aviation Week & Space Technology,* April 4, 1988, 46.

Hammer, Alexander R. "Home Front Helicopters," *New York Times,* May 23, 1971, F4.

"Helicopter Shuttle Saves Many at Las Vegas Fire," *Aviation Week & Space Technology,* December 1, 1980, 21.

"Helicopter Will Shuttle Passengers from Garage," *Wall Street Journal,* April 14, 1947, 6.

Hevesi, Dennis. "Police-Fire Feuds: Only in New York," *New York Times,* May 6, 1988, B3.

Hiro, Dilip. *The Longest War: The Iran-Iraq Military Conflict.* New York: Routledge, Chapman and Hall, 1991.

Horne, George. "Capital Studies Copter Service," *New York Times,* September 8, 1966, 73.

Hudson, Edward. "City to Consider Pan Am Heliport," *New York Times,* July 17, 1963, 26.

"Hunt for Oil, Gas Quickens All Across U.S.," *U.S. News & World Report,* September 15, 1980, 51.

Immel, Patrick. 2007. Personal communication. 10 January.

"Iraqi Tank Guns Stop Missile Helicopters," *Aviation Week & Space Technology,* November 24, 1980, 66.

"It Was Death, Absolute Death," *Time,* December 1, 1980, 34.

Jalali, Ali Ahmad, and Lester W. Grau. *The Other Side of the Mountain: Mujahideen Tactics in the Soviet-Afghan War.* Quantico, VA: U.S. Marine Corps Studies and Analysis Division, 1995.

Jarboe, Jan. "Flight for Your Life," *Texas Monthly,* March 1, 1992, 96.

Johnson, Ken. 2006. Interview by author. 19 March.

Kaiser, Charles. "Helicopter Flights Approved by Board," *New York Times,* January 21, 1977, 82.

Kendall, John. "Main Lesson of High-Rise Fire: Sprinklers Are Vital," *Los Angeles Times,* May 16, 1988, 1.

Khosa, Raspal S. "The Siachen Glacier Dispute," *Contemporary South Asia,* July 1999, 187.

Kinghorn, Spike. 2006. Interview by author. 22 March.

Klem, Thomas. "Los Angeles High-Rise Bank Fire," *Fire Journal*, May–June 1989, 72.

Klose, Kevin. "Heroics Amid Panic; Guests Helped Others to Roof to Await Copter," *Washington Post*, January 2, 1987, A1.

Kluckhorn, Frank L. "Boston Is Using Helicopters for Trip to Airport," *New York Times*, July 27, 1947, E6.

Larsen, Ron. "Coast Guard: Can You Assist at a Hotel Fire?" *Fire Command*, March 1987, 23.

Larson, Mel. 2002. Interview by author. 10 April.

Levin, Alan, and Kevin Johnson. "Air Ambulance Crashes Spur Safety Reviews," *USA Today*, January 14, 2005, A3.

Levy, Clifford J. "Fire Chiefs Assail Rescue After Bombing," *New York Times*, April 11, 1993, 28.

Lindsey, Robert. "Helicopters Aid Commuting," *New York Times*, October 16, 1969, 49.

"Los Angeles Studies Flying Bus for Airport Transit," *New York Times*, December 18, 1966, S19.

Mackby, Jenifer. "Helicopters After the Vietnam Era," *New York Times*, July 20, 1975, F1.

Maher, Marie Bartlett. *Flight for Life*. New York: Pocket Books, 1993.

McGirk, Jan. "Stand-Off in the Peaks of Kashmir," *The Independent*, July 1, 2005, 28.

McGowan, Jay. 2006. Interview by author. 23 February.

Meadows, Mike. "LAFD's Flying Firefighters," *Firehouse*, March 2001, 90.

Meier, Barry. "Air Ambulances Are Multiplying, and Costs Rise," *New York Times*, May 3, 2005, A1.

Middleton, Drew. "Tactics in Gulf War," *New York Times*, October 19, 1980, 12.

"Military Surplus Goods Fuel Bogus Parts Market," *Aviation Week & Space Technology*, March 1, 1993, 56.

National Transportation Safety Board. *Commercial Emergency Medical Service Helicopter Operations, Safety Study NTSB/SS-88-01*. Washington, D.C.: NTSB, 1988.

"NYA Predicts Mass Copter-Commuting," *Aviation Week*, December 8, 1952, 87.

Parrish, Roy L. "The MGM Grand Hotel Fire," *International Fire Chief*, January 1981, 12.

"Police Authorized to Buy Helicopter for $25,000," *New York Times*, August 11, 1948, 23.

"Port Authority Urges 'Heliports' for N.Y. City," *New York Times*, February 8, 1950, 12.

Proctor, Paul. "Aeromedical Aircraft Accidents Register Sharp Increase in 1987," *Aviation Week & Space Technology,* July 13, 1987, 55.

―――. "Cedar Logging Tests Pilot Skills," *Aviation Week & Space Technology,* April 22, 1996, 66.

Purdum, Todd S. "Race to Rescue: Police-Fire Feud Dates from the 30s," *New York Times,* June 9, 1988, B1.

Regan, Joe. "Firefighting Takes to the Air," *Firehouse,* June 1979, 11.

Rhodes, J. David, and Matt Moseley. "Atlanta Mill Fire and Helicopter Rescue," *Fire Engineering,* June 1999, 83.

Ruhl, Robert K. "The Pony Rides in Vegas," *Airman,* April 1981, 2.

Sharry, John A. "South America Burning," *Fire Journal,* July 1974, 23.

Sheehan, Edward R. F. "The Epidemic of Money," *New York Times,* November 14, 1976, 224.

"South Vietnamese Airman Defects with 6 Others to Thailand in Copter," *New York Times,* March 10, 1976, 2.

Sowa, Randall. 2007. Interviews and personal communication with author. January.

Stameisen, Gary. 2002. Interview by author. 15 February.

Stamm, William. "The SMS: Aerospace Technology for the Fire Service," *Fire Command,* August 1978, 51.

Steinbrunn, Robert. "Personal Journal: Amputation Weekend," *Hospital Aviation,* September 1985, 14.

Stevens, Charles W. "Sales of Helicopters Soar as Concerns Use Them to Speed Up Operations and Executives' Travel," *Wall Street Journal,* July 6, 1979, 28.

"Three Die in Amtrak Derailment," *Burlington Free Press,* July 8, 1984, 1.

Tinsley, Frank. "Copter Commuting: You'll Be Doing It Soon," *Collier's,* February 14, 1953, 7.

Tishchenko, Marat. 2005 and 2006. Personal communication. December.

Utz, Eugene. "Los Angeles—Helicopter Town of the Month," *American Helicopter,* September 1947, 19.

Venter, Al J. *The Chopper Boys: Helicopter Warfare in Africa.* London: Greenhill Books, 1994.

Walcott, John, and Tim Carrington. "Role Reversal: CIA Resisted Proposal to Give Afghan Rebels U.S. Stinger Missiles," *Wall Street Journal,* February 16, 1988, 1.

Walker, C. Lester. "Tomorrow's Helicopters," *Harper's,* May 1953, 28.

―――. "Age of the Whirling Wings," *New York Times,* September 16, 1956, SM10.

Walker, John. 2002. Interview by author. 7 February.

Wall, Robert. "MH-47 Crews Detail Conflict's Exploits, Woes," *Aviation Week & Space Technology*, April 15, 2002, 22.

Warner, Jack. "Hot Flames, Cool Heads: Heroic Drama Seemed Made for TV," *The Atlanta Journal*, April 19, 1999, A1.

West, Jim. "Adams: Look to the Rockies for a Challenge in Exploration," *Oil & Gas Journal*, November 16, 1981, 141.

Westenhof, Charles M. "Airpower and Political Culture," *Airpower Journal*, Winter 1997, 39.

Wheeler, David. "175 Rescue Workers Hear of Amtrak Derailment," *Burlington Free Press*, November 23, 1984, 5B.

Wiegner, Kathleen K., and Ellen Paris. "Here Come the Helicopters," *Forbes*, October 12, 1981, 132.

Willey, A. Elwood. "High-Rise Building Fire," *Fire Journal*, July 1972, 7.

Williams, Gurney. "Flying Fire Engine," *Popular Mechanics*, April 1979, 90.

Witkin, Richard. "Airlines Still Wait for Manufacturers to Produce a Practical Transport," *New York Times*, September 16, 1956, 30.

———. "Helicopter Landing Gear Blamed," *New York Times*, May 18, 1977, 47.

Wright, Chapin. "During Feud, Safety Hung in the Balance," *New York Times*, May 7, 1990, 1.

Chapter Thirteen

"Air Pegasus Forced to Close D.C. Heliport," *Airports*, July 30, 2002, 7.

"An Effort to Suppress Noise," *Forum*, April 1906, 552.

Barnes, Frank, 2005. Interview by author. 4 February.

Bender, Marylin. "Jet-Age Commuter Also a Family Man," *New York Times*, September 15, 1962, 16.

"Berkeley Council Rejects Copter Patrols by Police," *New York Times*, May 31, 1970, 3.

"Boston Helicopter Service Suspends Operations," *Wall Street Journal*, August 2, 1947, 2.

Butler, Robert. 2005. Interview by author. 7 February.

Caldeira, Teresa Pires do Rio. *City of Walls: Crime, Segregation, and Citizenship in São Paulo*. Berkeley, CA: University of California Press, 2000.

Caldwell, Earl. "Berkeley Faculty Urges Inquiry into 'Lawlessness' by Police," *New York Times*, May 24, 1969, 23.

Cerra, Frances. "Residents Complain of Noise by New Copter Shuttle," *New York Times*, November 11, 1982, B1.

Clausen, Meredith L. *The Pan Am Building and the Shattering of the Modernist Dream.* Cambridge, MA: MIT Press, 2004.

Cogan, Charles G. "Desert One and Its Disorders." *The Journal of Military History,* January 2003, 201.

Collitt, John. "Kidnapping Spreads Across Latin America," *Financial Times,* November 29, 2002, 11.

Conboy, Ken, and James Morrison. "The Quiet One," *Air Forces Monthly,* April 1998, 43.

"Copter Breaks Up Berkeley Crowd," *New York Times,* May 21, 1969, 1.

"Crash of Copter in '77 on Pan Am Building Was the Area's Worst," *New York Times,* April 19, 1979, B6.

Custis, Jon A. "Fire Force: Vertical Envelopment during the Rhodesian War," *Marine Corps Gazette,* March 2000, 48.

Cwerner, Saulo B. "Vertical Flight and Urban Mobilities: The Promise and Reality of Helicopter Travel," *Mobilities,* July 2006, 191.

Demarr, Jim. 2005. Interview by author. 10 October.

DeMeis, Rick. "Quieting Black Hawk Down," *EDN,* August 8, 2002, 26.

Dethman, Leigh. "Luck, Volunteers, Movie Star Played Roles in Scout Rescue," *Deseret Morning News,* June 22, 2005, A13.

Downie, Andrew. "A Stone's Throw from Poverty, Brazil's Daslu Glitters," *Christian Science Monitor,* July 12, 2005, 4.

Dubin, Zan. "The Sky's the Limit: Helicopters Have a Means for Some to Rise Above Gridlock," *Los Angeles Times,* November 15, 1990, 1.

Ebersole, Mike. 2005. Interview by author. 6 February.

Esler, David. "Helicopter Operations in Temporary Landing Areas," *Business & Commercial Aviation,* February 2003, 56.

"Eurocopter's Electrical Flap Control Breakthrough," *Interavia,* Autumn 2005, 10.

Faiola, Anthony. "For the Elite, a High Road—Commuting via Helicopter," *Washington Post,* June 11, 2002, 24A.

Farrelly, Paul. "Cartel That Conceals Its Cutting Edge," *The Observer,* March 4, 2001, 4.

Federal Aviation Administration. *Report to Congress: Nonmilitary Helicopter Urban Noise Study.* Washington: Federal Aviation Administration, 2004.

"Flies from Boston Roof," *New York Times,* April 17, 1947, 55.

"Flights of Folly," *The Nation,* May 28, 1977, 644.

Ford, Harrison. 2006. Interviews and personal communication with author. February.

Grant, Keith. 2005. Interview by author. 15 October.

Green, Richard. 2005. Interview by author. 6 February.

Grissett, Sheila. "Ochsner Neighbors Blast Helipad," *Times-Picayune*, February 17, 2004, 1.

Heinl, Robert D. "The Woman Who Stopped Noise," *Ladies' Home Journal*, April 1908, 19.

Held, Joy. 2005. Interview by author. 14 October.

"Heliport," *The New Yorker*, January 1, 1966, 19.

Hinds, Michael deCourcy. "By Copter to the Airports, Far Above the Potholes," *New York Times*, May 31, 1981, 1.

Hinkle, Rick. 2005. Interview by author. 10 October.

Hodson, Mark. "Washington DC by Helicopter," *Sunday Times*, July 23, 2000, 4.

"Hover Bother," *Flight International*, September 23, 2003, 3.

"How Would You Respond to the Helicopter/Heliport Opponents?" *Eastern Region Helicopter Council News*, Summer 2005, 10.

Hudson, Edward. "Mayor Endorses Heliport Site at East River and 61st Street," *New York Times*, January 3, 1968, 1.

———. "Heliport Opened Atop Skyscraper," *New York Times*, December 22, 1965, 26.

———. "Helicopter 'Taxi Stand' for Busy People Opens in Midtown on East River," *New York Times*, November 5, 1968, 49.

———. "Pan Am Neighbors Attack Heliport," *New York Times*, May 22, 1963, 58.

———. "Helicopter Line Here Is Hopeful," *New York Times*, January 27, 1965, 70.

Jumpei, Marcio. "Brazil's Capital, South America's Largest City, Has Embraced the Helicopter Like No Other," *Business & Commercial Aviation*, March 1, 2005, 58.

Kihss, Peter. "Helicopter Hijacked to Pan Am Building," *New York Times*, May 24, 1974, 69.

Kindleberger, Richard. "Owner Closes Boston's Only Commercial Airport," *Boston Globe*, April 3, 1999, B4.

Kluckhorn, Frank. "Boston Is Using Helicopters for Trip to Airport," *New York Times*, July 27, 1947, E6.

Koklanaris, Maria. "Heliport Plan Stirs Opposition," *Washington Post*, March 15, 1990, J1.

Lambert, Bruce. "Heliport: New Lease and Quiet," *New York Times*, October 8, 1995, CY6.

———. "Roar of Helicopters Brings a Whirl of Residents' Protests," *New York Times*, January 7, 1996, CY6.

Larsen, Dave. "Ford, Flight & Fame," *Dayton Daily News*, July 13, 2003, F1.

Lewan, Todd. "Gigayachts Taking Industry by Storm," *Journal-Gazette* (Fort Wayne, IN), August 14, 2005, 1D.

Lopez, Steve. "One Way to Get Closer to Heaven: Helicopters," *Los Angeles Times,* July 18, 2001, B1.

"Los Angeles Helicopter Service Resuming After Second Crash," *New York Times,* August 20, 1968, 82.

Malone, Pat. "Whirly Commuters Are Go," *Sunday Times,* December 11, 2005, 6.

McCulloch, Campbell. "Taxis of the Air," *McClure's Magazine,* June 1919, 27.

McGarry, T. W. "Airport Foes Arming for War Over Helicopter Shuttle," *Los Angeles Times,* October 20, 1987, 8.

McSkimming, Jen. 2005. Interview by author. 6 February.

Milhorn, Mike. 2005. Interview by author. 7 February.

Moses, Robert. *Public Works: A Dangerous Trade.* New York: McGraw-Hill, 1970.

"Mrs. Isaac L. Rice, Foe of Noise, Dies," *New York Times,* November 5, 1929, 28.

Murray, Kathleen. "Sky Driving: Commuters Take to the Air," *Orange County Register,* September 8, 1990, F1.

Naughton, Keith. "The Fast and the Luxurious," *Newsweek,* January 13, 2003, 40.

"New York's Heliports," *Business & Commercial Aviation,* April 2002, 86.

O'Donnell, Michelle. "Boon or Plague, 10 More Years of Whup-Whup," *New York Times,* March 3, 2002, 14.

Oates, Mary Louise. "Madonna, Penn: It's a Glitzy Wedding," *Los Angeles Times,* August 17, 1985, 1.

Pasternak, Judy. "Ex-Guru Seeks to Expand His Heavenly Rights," *Los Angeles Times,* April 11, 1985, 1.

———. "Maharaji Denied in Bid to Triple Copter Use," *Los Angeles Times,* July 7, 1985, 1.

Pillsbury, Fred. "The Use of Helicopters Is Soaring," *Boston Globe,* August 22, 1984, 1.

Posey, Carl A. "São Paulo Traffic Report," *Air & Space,* October–November 2002, 48.

"President's Helicopters Emulate 1911 Landing on White House Lawn," *New York Times,* June 1, 1957, 38.

Profico, John. 2005. Interview by author. 10 October.

Ramirez, Anthony. "The Chopper Blocker," *New York Times,* March 2, 1997, 13.

———. "Helicopters Won't Cut Back Without a Fight," *New York Times,* September 22, 1996, CY10.

Rice, Julia. "Our Most Abused Sense—The Sense of Hearing," *Forum,* April 1907, 559.

Robinson, Frank. 2005. Interview by author. 6 February.

Rorabaugh, W. J. "The Battle of People's Park," *San Francisco Chronicle,* May 14, 1989, 7.

Schanberg, Sydney H. "After Year and a Half, Copter Critics Are Quieter," *New York Times,* July 23, 1967, 1.

Seabrook, John. "The Slow Lane: Can Anyone Solve the Problem of Traffic?" *The New Yorker,* September 2, 2002, 120.

Steinhauer, Jennifer. "No Big Deal, Mayor Says of His Helicopter Flying," *New York Times,* February 26, 2002, B4.

Stevens, Charles W. "And Here Comes a Chopper, to Make Neighbors See Red," *Wall Street Journal,* July 31, 1990, A1.

Stewart, James B. "Spend! Spend! Spend! Where Did Tyco's Money Go?" *The New Yorker,* February 17, 2003, 132.

Stockbridge, Frank Parker. "The War on Noise," *New McClure's,* December 1928, 66.

Svard, Trygve. 2004. Interview by author. 6 June.

"The Bird Man of Torrance," *Forbes,* April 15, 1991, 64.

"The Children's Hospital Branch of the Society for the Suppression of Unnecessary Noise," *Forum,* April 19, 1908, 560.

"The Doomsday Blueprints," *Time,* August 10, 1992, 32.

Thomas, Rhys. 2006. Interviews by author. June.

Thorsrud, Derek. 2005. Interview by author. 19 December.

Vasquez, Liliana. 2006. Personal communication. 24 February.

Wagner, Patricia. 2005 and 2006. Interviews and personal communication with author. October and February.

Weisel, Al. "Half a Copter," *Fortune,* February 21, 2000, 310.

Williams, John D. "Opposition to Helicopters Is Increasing in Many Cities, and the Industry Is Fearful," *New York Times,* August 25, 1983, 46.

Witkin, Richard. "Standard in European Cities, They Are Denied Good Locations Here," *New York Times,* April 24, 1955, X33.

Woodyard, Chris. "Big Dreams for Small Choppers Paid Off," *USA Today,* September 12, 2005, B7.

Zwingle, Erla. "São Paulo, Brazil: World's Third Largest City," *National Geographic,* November 2002, 72.

Chapter Fourteen

Adelman, Kenneth. 2005. Personal communications. August.
"Air America: Hughes 500," at http://www.utdallas.edu/library/collections/speccoll/Leeker/500.pdf (Accessed July 29, 2006).
Bannon, Lisa. "In T.V. Chopper War, News is Sometimes a Trivial Pursuit," *Wall Street Journal,* June 4, 1997, A1.
Bart, Peter. "2,000 Troops Enter Los Angeles on Third Day of Negro Rioting," *New York Times,* August 14, 1965, 1.
"Bird's-Eye View," *Time,* August 4, 1958, 51.
"Camper Found in Time for Kidney Transplant," *Chicago Sun-Times,* March 27, 1989, 42.
Cannon, Lou. "Worlds Collide at Florence and Normandie," *Washington Post,* January 26, 1998, A14.
Chavez, Stephanie. "Trucker Looks Back on Los Angeles Riots," *Houston Chronicle,* April 26, 2002, 4.
Conboy, Ken, and James Morrison. "The Quiet One," *Air Forces Monthly,* April 1998, 43.
Davies, Lawrence E. "Berkeley Police Seeking Copters," *New York Times,* February 10, 1970, 24.
Elber, Lynn. "Newsman, 75, Shares TV's Five-Decade History," *Tulsa World,* July 15, 1998, 4.
Ellis, Bill. 2006. Personal communication. 8 February.
———. *Raising the Devil: Satanism, New Religions, and the Media.* Lexington, KY: University Press of Kentucky, 1998.
Guthrie, C. Robert, and Los Angeles County (Calif.) Sheriff's Dept. *Project Sky Knight: A Demonstration in Aerial Surveillance and Crime Control. Final Report to Office of Law Enforcement Assistance.* Washington: Office of Law Enforcement Assistance, U.S. Dept. of Justice, 1968.
Harmon, Dave. "Invasion, South Texas: Army Exercises Make Black Helicopters the Talk of Town," *Austin American-Statesman,* April 17, 1999, A1.
"Hot Shots," *People Weekly,* September 12, 1994, 97.
"In the Sky, on the Air," *Los Angeles Daily News,* March 5, 1997, L4.
Kisseloff, Jeff. *The Box: An Oral History of Television.* New York: Viking, 1995.
Leyvas, Gil. 2005. Interview by author. 4 February.
Lindsey, Robert. "Police Send Up Copters in Fight on Urban Crime," *New York Times,* December 9, 1970, 37.
Lineberry, Gary. 2004. Interview by author. 11 December.

Littleton, Cynthia. "KTLA, the West's Golden Station," *Broadcasting & Cable,* April 28, 1997, 26.

"Los Angeles TV Gives Viewers Riot Coverage," *New York Times,* August 15, 1965, 79.

McDougal, Dennis. "In the Eye of the Storm: Chopper Pilots—High Visibility Heroes of the Air," *Los Angeles Times,* February 14, 1992, 1.

Melton, Mary. "If It Speeds, It Leads," *Los Angeles Magazine,* February 2003, 50.

Newton, Jim, and Beth Shuster. "The North Hollywood Shootout," *Los Angeles Times,* March 4, 1997, 1.

Oldfield, Tom. 2004. Interview by author. 11 December.

Pipes, Daniel. *Conspiracy: How the Paranoid Style Flourishes and Where It Comes From.* New York: Free Press, 1997.

Pool, Bob. "Serene Hilltop Marks Site of Landmark Disaster," *Los Angeles Times,* December 11, 2003, B2.

Purdum, Todd S. "Vigilant Eyes Fill Skies Over Los Angeles," *New York Times,* March 18, 1998, A1.

Richmond, Ray. "Copter Pilots Keep TV News on Top of Story," *Orange County Register,* February 4, 1991, FO3.

Rosenberg, Howard. "The Russian Roulette of Live News Coverage," *Los Angeles Times,* May 2, 1998, 1.

Rubinkam, Michael. "Philadelphia Police, Mayor Under Fire for Video-taped Beating," *Times Union* (Albany, NY), July 14, 2000, A3.

Schulberg, Pete. "Bright Images, Blurry Ethics," *The Oregonian,* October 6, 1996, E1.

Skelton, Kevin. 2005. Personal communication. 21 December.

Tur, Robert. 2006. Interviews by author. August.

Ulman, Neil. "More Radio Stations Use 'Copters to Spot Road Jams for Drivers," *Wall Street Journal,* December 24, 1964, 1.

Weinraub, Bernard. "TV News Displays Air Power in Chase," *New York Times,* June 20, 1994, A12.

Welk, Larry. 2005. Interviews by author. February.

White, Garrett. "Night Riders," *Los Angeles Magazine,* November 1997, 108.

Wolfe, Dan. 2004. Interview by author. 11 December.

Wright, Jeff. 2005. Interviews and personal communication with author. October–November.

Chapter Fifteen

Ballou, Justin G. (Guy). 2006. Personal communications. March.

Beckey, Fred. *Mount McKinley: Icy Crown of North America.* Seattle: The Mountaineers, 1993.

Bowden, Mark. *Black Hawk Down: A Story of Modern War.* New York: Atlantic Monthly Press, 1999.

Brown, Wayne. 2006. Interview by author. 17 February.

Calvery, Donald. 2006. Interview by author. 6 March.

"Copter Saves 3rd Climber," *Chicago Sunday Times,* May 22, 1960, 1.

Crews, Paul. "Accident on Mount McKinley," *Summit,* August 1960, 2.

Curtis, Ian G. S. "Changing Helicopters for a Rapidly Changing World," *Defense & Foreign Affairs Strategic Policy,* August 1997, 7.

Day, John S. "The Mountain That Nearly Killed Me," *Saturday Evening Post,* November 26, 1960, 36.

Erickson, Scott. 2005. Interview by author. 16 December.

Farabee, Charles. *Death, Daring and Disaster.* Emeryville, CA: Roberts Rinehart Publishing, 1998.

Feron, James. "Israelis Seize 7-Ton Radar and Airlift It Out of Egypt," *New York Times,* January 3, 1970, 1.

Gadbois, Chris. 2005. Interview by author. 7 February.

Gant, Dale. 2005. Interview by author. 29 December.

Gray, Sidney J. "The 160th SOAR: 20 Years of Army Special-Operations Aviation," *Special Warfare,* Summer 2001, 6.

Halloran, Richard. "Secret U.S. Army Unit Had Role in Raid in Gulf," *New York Times,* September 24, 1987, A12.

"Helicopter Saves Two Hurt on Peak," *New York Times,* May 22, 1960, 43.

Hepler, Robert. 2006. Personal communications. June–July.

Hewson, Harry J. "Light/Attack Helicopter Operations in the Three Block War," *Marine Corps Gazette,* April 1999, 25.

Humes, Edward. "Helicopter Crashes: 134 Lives Lost Since Pilots Began Using Goggles," *Orange County Register,* December 4, 1988, 14.

Hurst, Arlo. 2006. Interview by author. 9 January.

Klem, Thomas J. "Los Angeles High-Rise Bank Fire," *Fire Journal,* May–June 1989, 72.

Luckett, Lincoln. 2005 and 2006. Personal communications. June–April.

Lynch, Brian. 2006. Interview by author. 9 January.

Maier, Karl. 2006. Interview by author. 9 January.

"Men Against the Mountain," *Time,* May 30, 1960, 1.

Milani, Andy. "Evolution of the 3-160th SOAR Through Desert Storm," *Special Warfare,* Summer 2001, 14.

Mills, Hugh. 2007. Interview by author. 19 March.
Moon, Ken. 2006. Interview by author. 27 March.
O'Brien, Clifton. 2006. Interview by author. 9 January.
Phillips, Ken. 2005. Interview by author. 6 February.
"Plane Rescues Stricken Woman from Mt. McKinley," *New York Times,* May 21, 1960, 1.
Prox, John. 2005. Interview by author. 3 October.
Raaz, Dana. 2005. Interview by author. 18 February.
Rudert, Dan. 2005. Interviews by author. May.
Sokalski, Walt. "Learning to SOAR," *Soldiers,* December 1998, 28.
Tamburro, Chuck. 2007. Interview by author. February 22.
"Task Force 160," at http://www.nightstalkers.com (Accessed November 15, 2006).
Thomas, Timothy L. "Air Operations in Low Intensity Conflict: The Case of Chechnya," *Airpower Journal,* Winter 1997, 51.
"Times Says Pentagon Formed 'Secret' Forces," *San Francisco Chronicle,* August 24, 1987, 9.
Waterman, Jonathan. *Surviving Denali: A Study of Accidents on Mount McKinley 1903–1990.* New York: American Alpine Club Press, 1983.
Whittaker, Lou. *A Life on the Edge.* Seattle: The Mountaineers, 1999.
Whittaker, Lou, and Andrea Gabbard. *Lou Whittaker: Memoirs of a Mountain Guide.* Seattle: The Mountaineers, 1994.
Willenbacher, Samantha. 2005. Interview by author. 12 August.

Chapter Sixteen

Ames, W. B. "Arizona Pioneer," *American Helicopter,* August 1947, 24.
Baker, Tony. 2005. Interviews by author. June.
Bodo, Sandor. "The Age of Air Lost Civilization," *Providence Journal,* January 28, 1990, M18.
Bol, Tom. "Denali Patrol," *Alaska,* May–June 2002, 24.
Brass, Eric H., Roy Braybrook, and John Burley. "The Tank Killers," *Armada International,* December 1998–January 1999, 21.
Brookings Institution. "Iraq Index: Tracking Variables of Reconstruction & Security in Post-Saddam Iraq," at http://www.brookings.edu/iraqindex (Accessed February 1, 2007).
Chamberlain, Gethin. "Uday and Qusay Die in Gun Battle Following Tip-Off," *The Scotsman,* July 23, 2003, 2.
Christenson, Sig. "Shot Down," *San Antonio Express-News,* March 22, 2004, at http://www.mysanantonio.com/news/military/stories/

MYSA 22.01A. longbow_2_0322.cfs7fd6.html (Accessed November 10, 2006).

Clark, Fred. 2007. Interviewed by author, April 25.

"Climbers Rode Out on 'God Ring,'" *Cincinnati Post,* June 23, 1998, 2A.

Corr, Patrick. 2005. Interview by author. 12 August.

Dubé, John. 2005. Interviews by author. June.

Dyer, Craig. 2007. Personal communication. 14 February.

Fulghum, David A., and Robert Wall. "Israel Refocuses on Urban Warfare," *Aviation Week & Space Technology,* May 13, 2002, 24.

Harris, Francis. "'Aerial Bombs' Threaten U.S. Helicopters in Iraq," *The Daily Telegraph* (London), January 18, 2006, O14.

Hilsum, Lindsey. "Iraq Hails David Who Felled Flying Goliath," *The Daily Telegraph,* March 25, 2003, EO3.

Hoffman, Carl. "Higher Calling," *Air & Space Smithsonian,* June–July 1998, 24.

Hood, Jim. 2005. Interview by author. 8 June.

Kaplan, Lawrence F. "The Airport Road," *Wall Street Journal,* January 27, 2005, A12.

Kramer, Mark. "The Perils of Counterinsurgency; Russia's War in Chechnya," *International Security,* Winter 2004–2005, 5.

Kreutzer, Dave. 2005. Interview by author. 8 June.

Kunz, Mike. 2005. Interviews and personal communication with author. April–June.

Linklater, Eric. *Private Angelo.* New York: Macmillan, 1946.

McDowell, Edwin. "At Trade Center Deck, Views Are Lofty, as Are the Prices," *New York Times,* April 11, 1997, B8.

McPhee, Michele, and John Marzulli. "WTC Doors Locked—Rudy Says Copter Rescue Would Have Been Too Risky," *New York Daily News,* October 24, 2001, 7.

Merle, Renae. "Low-Tech Grenades a Danger to Helicopters," *Washington Post,* November 18, 2003, A17.

Merritt, Larry. 2002. Interview by author and personal communication. April.

Middleton, Drew. "Army Eyes Division with Antitank Copters as an Answer to Russian Armor in Europe," *New York Times,* May 1, 1972, 20.

Morris, Robert R. "Hydraulic Forcible Entry Tools," *WNYF,* Third Quarter, 2001, 7.

Musquere, Anne. "High Times for Helo Makers," *Interavia,* Spring 2006, 14.

National Commission on Terrorist Attacks Upon the United States. *The 9/11 Commission Report.* New York: W. W. Norton.

Newman, Richard J. "Ambush at Najaf." *Air Force Magazine,* October 2003, 60.

Paltrow, Scot J., and Queena Sook Kim. "No Escape: Could Helicopters Have Saved People from the Trade Center?" *Wall Street Journal,* October 23, 2001, A1.

Phillips, Natalie. "High Rescue Costs Add Up," *Anchorage Daily News,* August 23, 1998, A1.

Porco, Peter. "Nonfatal Attraction," *Anchorage Daily News,* May 3, 2005, A1.

Powell, Colin. 2006. Interview by author. 23 March.

Rich, Rob. 2005. Interview by author. 6 February.

Rodway, George. "Paul Crews 'Accident on Mount McKinley'—A Commentary," *Wilderness and Environmental Medicine,* 2003, 33.

Roug, Louise. "Troops Have a Nervous Ride to Nighttime Raid," *Los Angeles Times,* November 7, 2005, A8.

Schmitt, Eric. "Iraq Rebels Seen Using More Skill to Down Copters," *New York Times,* January 18, 2004, 1.

Schneider, Larry. 2002. Interview by author. 15 March.

Scott, Brian. "Enthusiasts Entertained by Son of Aviation Pioneer," *The Beacon* (Gander, Newfoundland, Canada), February 26, 2007, at http://www.transatlanticflightplay.com/files/Beacon Article Gala.pdf (Accessed March 15, 2007).

Steele, David. "Locked Doors to Roof Cost Lives," *The Herald,* October 25, 2001, 14.

Wall, Robert. "IAF to Modify Helos to Fight in Cities," *Aviation Week & Space Technology,* May 13, 2002, 27.

Wilson, Jim. "Weapons of the Insurgents," *Popular Mechanics,* March 2004, 64.

An extended bibliography is available online at www.thegod machine.us.

INDEX

Page numbers of illustrations appear in italics.

80–83; KD-1, 85; lack of success, 86, 87, 87n; mail route, 85–86; PCA-2, 79, 82; pilots, 86; Pitcairn OP-1, 78; roadworthy, 84, 84n; safety issues, 86; uses, 78–79, 84, 85–86. *See also* Cierva, Juan de la
Autogiro Company of America, 79, 87
Avolater, 103n

Baden Baden, 116, 116n
Bading, Helga, 246, 265
Baker, Tony, 273
Baldwin, Tom, 52
Ballauer, Alb, 143n
balloons, 38, 39, 42, 275; danger of, 48–49; engine-powered, 44; first aerial photography, 40; grand voyage, 48; manned flights, 27–28; military applications, 59–60; Nader and, 40–41, 48–49; propulsion and, 43–44; rescues, 111n; Santos-Dumont's small powered balloons, 49; steam engines and, 44; vulnerability, 60; wartime evacuation, 181n; Wise's *Atlantic*, 43–44; World War I, 59
Ballou, Justin G. "Guy," 251–53
Bancroft, Arthur, 149
Bane, Thurman H., 63, 64, 97
Barnecott, Anne, 212
Barnum, P. T., 37n
Beame, Abraham, 212
Beau de Rochas, Alphonse, 46
Bell, Alexander Graham, 52, 55, 57
Bell, Derek, 235
Bell, Larry, 118, 118n, 126
Bell Aircraft, 222
Bell Aircraft/Bell Helicopter, 118, 195; Arthur Young and, 117–20; experiments with armed helicopters, 152–53; financial problems, 146
Bell helicopters, 115, 132n; AH-1 Cobra gunship, 174; AH-1J Sea Cobra gunship, 195, 196; HSL, 119;

JetRanger, 2–3, *32*, 199, 200, 232, 243–44, 245; Long Rangers, 198; military evacuation and battlefield observation, 147; Model 30, 118; Model 47 and variations, 119, *119*, 120, 135–36, 137, 140, 152–53, 189n, 270; OH-13 Sioux, 161; 212, 198; UH-1 Iroquois (see Huey helicopters)
Bendix Whirlaway, 138, 146
Bergenac, Bill, 178n
Berliner, Emile, 53, 57–58, 66
Berliner, Henry, 66–67
Berliner No. 5 helicopter, 66–67
Berry, Sidney, 186
Bienvenu, François, 27–29, 30
Bingham, Hiram, 77
Birds of Prey (film), 272n
Black Hawk. *See* Sikorsky helicopters
black ops and night-flying operations, 224–30; black-helicopter cult, 228–30, 228n, 229n; Kingsville, Texas, 228
Bleriot, Louis, 94
Bloomberg, Michael, 214
Blue Thunder (film), 229, 239
Boeing: AH-6/MH-6 Little Bird, 224, 256–57, 258, 259; CH-46 Sea Knight helicopter, 36, 190, *191*, 209, 212; CH-47 (*see* Chinook helicopter); V-22 Osprey (with Bell Helicopter), 256
Bölkow helicopter, 198
boomerangs and throwing sticks, 6–7
Bossi, Enea, 115
Boulet, Jean, 71n
Brantly Helicopters, 115
Brazil: São Paulo rooftop rescues, 203–4, 203n; São Paulo helicopters, 214–15
Bréguet, Louis-Charles, 54–55, 75n, 90, 100, 115
Brennan, Louis, 7–8, 8n, 68, 117
Brequet, Louis-Charles and Jacques, 53
Bright Shining Lie (Sheehan), 180

About the Author

JAMES R. CHILES trained in a two-seater helicopter as part of his research for this book. His first book, *Inviting Disaster,* was named a Best Book of 2001 by Amazon.com and made into a four-part series for the History Channel. He lives in Minnesota. He can be reached via www.thegodmachine.us.